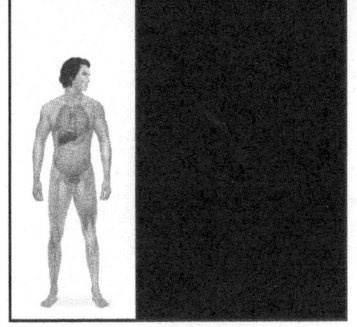

Student Workbook

to Accompany

Anatomy, Physiology & Disease
An Interactive Journey
for Health Professionals
Fourth Edition

Bruce J. Colbert
Director of Allied Health
University of Pittsburgh at Johnstown

Jeff Ankney
Director of Clinical Education, Respiratory Care Program
University of Pittsburgh at Johnstown

Karen T. Lee
Associate Professor of Biology
University of Pittsburgh at Johnstown

Boston • Columbus • Indianapolis • New York • San Francisco • Upper Saddle River
Amsterdam • Cape Town • Dubai • London • Madrid • Milan • Munich • Paris • Montréal • Toronto
Delhi • Mexico City • São Paulo • Sydney • Hong Kong • Seoul • Singapore • Taipei • Tokyo

330 Hudson Street, New York, NY 10013

ISBN 10: 0-13-453127-2
ISBN 13: 978-0-13-453127-4

Pearson

5 2021

CONTENTS

PREFACE

This workbook is designed to accompany the textbook *Anatomy, Physiology & Disease: An Interactive Journey for Health Professionals, Fourth Edition*. Many of its features will allow you to work at your own pace, helping you to evaluate your progress throughout the course. This workbook features a number of different ways to assess your progress as you journey through the text. Types of questions include:

Chapter Outline and Chapter Summary—These sections provide an at-a-glance refresher of the key topics covered in the chapter.

Medical Terminology Review—This section provides a review of key disease-related terms presented in the chapter.

Matching—In this section, there are four exercises per chapter in which you are asked to match chapter terminology, concepts, and disorders with appropriate definitions, descriptions, symptoms, or treatments.

Multiple Choice and Fill in the Blank—These sections provide an assessment to help test your knowledge of the chapter material.

Labeling—In some chapters, you are asked to identify structures and label and/or color a blank image with the appropriate structures. These diagrams correspond to those found in the textbook.

Case Studies and Short Answer—These sections help you to think critically and apply the chapter knowledge.

Learning Activities—These activities include engaging team projects, learning game ideas, and research topics.

We hope this workbook helps make your journey through anatomy and physiology an enjoyable one.

ANATOMY, PHYSIOLOGY, & DISEASE: LEARNING THE LANGUAGE

 ## CHAPTER SUMMARY

Anatomy is the study of the *structure* of living things—in this case, the human body. The study of structures that can be seen with the naked eye is called macroscopic, or gross, anatomy, whereas the study of structures that can be seen only with magnification is called microscopic, or fine, anatomy. Physiology is the study of the *function* of living things. Often the best way to understand how living things work is to study the anatomy and physiology at the same time, paying close attention to the relationship between the two.

When the body fails to function normally, a disease is present. A professional who studies the characteristics, causes, and effects of disease is called a pathologist. An epidemiologist studies the occurrence, distribution, and control of disease. When a person develops a disease, the pathologist and the epidemiologist will both be interested in determining the cause (etiology) of the disease. Diseases with undetermined causes are called idiopathic, whereas a disease acquired in a health care setting is called a nosocomial infection. Epidemiologists are particularly interested in communicable diseases, which can be spread. If the disease spreads from person to person, it is said to be contagious. Whether the disease is communicable or actually contagious, the distribution is very important in understanding and controlling the disease. Diseases may be endemic (native to a certain area), epidemic (occurring in large numbers in a certain area), or pandemic (common worldwide).

In order to understand anatomy, physiology, and/or disease, you have to understand the language of science and medicine. Just as in all professions, a set of terms is commonly used among medical professionals. These terms are typically some combination of word roots/combining forms, prefixes, and suffixes. If you know a few key roots, prefixes, and suffixes, you can figure out many medical terms. The word root is the basic structure. Prefixes (before the word root) and suffixes (after the root) are used to modify the root, forming dozens of terms from a single root. In addition, many medical terms have standard abbreviations that are used so commonly that you must be familiar with them to function as a medical professional. Medical and scientific measurements are also different from the measurements used on a day-to-day basis. Instead of the U.S. Customary System of feet, inches, pounds, and cups, medical and scientific measurements use the metric system, which is based on the number 10. Medical measurements are in meters, centimeters, grams, and milliliters. Once you get used to the metric system you will find it is actually much easier to use than the U.S. Customary System.

Two physiological concepts will keep popping up again and again as you move from system to system in the body: metabolism and homeostasis. Metabolism is all the reactions performed in all the parts of the body. Metabolism can be broadly divided into two types of reactions. Anabolic reactions (anabolism) build new molecules, whereas catabolic reactions (catabolism) destroy molecules to make raw material for anabolism. Homeostasis is the body's ability to regulate body chemistry within a narrow favorable range. Whenever a body parameter gets out of range, homeostasis brings the value back into normal range. The process of bringing the value back into range is called negative feedback. Negative feedback always counteracts a change. The body also has some positive feedback mechanisms. Positive feedback (a vicious cycle) exacerbates or enhances a change away from normal. Positive feedback is not considered to be part of homeostasis because positive feedback does not counteract a change, but rather makes it worse.

When a patient develops or acquires a disease, the body ceases to function normally. A disease is always accompanied by changes in body functions. These changes can be either signs (objective, measurable changes) or symptoms (somewhat subjective changes perceived by the patient). Rapid pulse, for example, is a sign because it can be objectively measured. Pain is a symptom because the perceived level of pain differs from patient to patient. A group of signs and symptoms related to a disease is often called a syndrome.

Signs and symptoms are important because they are used by medical professionals to diagnose (identify) the disease. The severity of symptoms may also aid in determining the prognosis (predicted outcome) of the disease. Diseases that result in death are called terminal diseases.

The body is able to defend against many diseases. The skin and other membranes keep many infections out of the body. If an infectious agent does get into the body, the immune response can often destroy it. Immune response is often accompanied by an inflammatory response, a necessary part of tissue repair. Inflammatory response is recognizable by the familiar symptoms of redness, heat, swelling, and pain at the site of injury or infection. Sometimes the immune system reacts to harmless substances and causes tissue damage. This response is called an allergy.

The body can become infected in many ways. The various ways are called routes of transmission. Vector-borne transmission involves the transmission of infection from an insect or other animal to a human. A biological vector transmits the pathogen from its body to yours, often when it bites you. A mechanical vector has the pathogen outside its body and transfers it to your skin, food, or water. Contact transmission can occur via direct or indirect contact. Direct contact involves the transfer of infected fluid into your body through a wound. Indirect contact occurs when disease is spread because you touch something that was touched by somebody carrying the infection. Common vehicle transmission occurs when people consume food, water, or medicine contaminated with the infectious agent. Airborne transmission occurs when a pathogen in the air, usually in droplets of saliva, is inhaled.

To prevent the spread of infection in health care settings and break the cycle of infection, a set of standard precautions should always be used during patient contact. The precautions, which range from hand washing to gloves to masks, gowns and protective eyewear, become more extensive as the possibility of infection from contact with body fluids increases. For example, when talking to a patient, only hand washing is necessary, but during surgery, full precautions must be taken including gowns, gloves, masks, and eye protection.

CHAPTER OUTLINE

I. What is anatomy and physiology?
   A. Anatomy
      1. Microscopic
      2. Macroscopic
   B. Physiology

II. What is disease?
   A. Pathology
   B. Terms
      1. Etiology
      2. Idiopathic
      3. Nosocomial
   C. Communicable vs. contagious
   D. Epidemiology
      1. Endemic
      2. Epidemic
      3. Pandemic

III. The language
   A. Medical terminology
      1. Word roots/combining forms
      2. Prefixes
      3. Suffixes
   B. Abbreviations
   C. The metric system

IV. Anatomy and physiology concepts
   A. Metabolism
      1. Anabolism
      2. Catabolism
   B. Homeostasis
      1. Negative feedback
   C. Positive feedback

V. Disease concepts
   A. Signs and symptoms
   B. Syndrome
   C. Diagnosis, prognosis, chief complaint
   D. Relapse, remission
   E. Mortality and morbidity

VI. The body's defense system and disease transmission
    A. Barriers
        1. Immune response
        2. Inflammatory response
    B. Routes of transmission
        1. Vector borne
           a. Biological vector
           b. Mechanical vector
        2. Contact transmission
           a. Direct
           b. Indirect
        3. Common vehicle
        4. Airborne
    C. Standard precautions

MEDICAL TERMINOLOGY REVIEW

Define the following terms.

1. Pathology: _____

2. Epidemiology: _____

3. Etiology: _____

4. Idiopathic: _____

5. Diagnosis: _____

6. Prognosis: _____

7. Syndrome: _____

8. Signs: _____

9. Symptoms: _____

10. Standard Precautions: _____

MULTIPLE CHOICE

Circle the letter of the correct answer.

1. The study of structure is called:
   a. pathology.
   b. anatomy.
   c. physiology.
   d. cytology.

2. The study of tissue structure and function is called:
   a. cytology.
   b. dermatology.
   c. histology.
   d. anatomy.

3. The study of function is called:
   a. physiology.
   b. anatomy.
   c. pathology.
   d. cytology.

4. The process of assessing the overall size and scarring pattern of the liver uses:
   a. macroscopic anatomy.
   b. microscopic physiology.
   c. macroscopic cytology.
   d. microscopic histology.

5. This system of measurement is also called the International System.
   a. USCS
   b. British Imperial
   c. English
   d. Metric

6. The science and study of the causes of diseases and their modes of operation is:
   a. anatomy.
   b. hepatology.
   c. diseasology.
   d. pathology.

7. A forecast of the probable course or outcome of a disease is a:
   a. sign.
   b. prognosis.
   c. diagnosis.
   d. symptom.

8. All chemical operations going on within the body are collectively known as:
   a. metabolism.
   b. homeostasis.
   c. syndrome.
   d. pathology.

9. Which system of measurement is used in places like the United Kingdom, Australia, and Canada?
   a. British Imperial
   b. Metric
   c. USCS
   d. Both a and c

10. In the English system, which is based on the British Imperial system, volume is expressed in:
    a. milliliters.
    b. kilograms.
    c. pints.
    d. cubic centimeters.

11. The foundation of a word is its:
    a. prefix.
    b. root.
    c. suffix.
    d. etiology.

12. In the U.S. Customary System, distance is expressed in:
    a. feet.
    b. kilograms.
    c. centimeters.
    d. liters.

13. In the metric system, weight is measured in:
    a. pounds.
    b. liters.
    c. ounces.
    d. kilograms.

14. Any abnormality indicative of disease and objectively discoverable on examination of a patient is a:
    a. syndrome.
    b. symptom.
    c. sign.
    d. septic.

15. The determination or identification of the nature of a disease, injury, or congenital defect is a:
    a. sign.
    b. syndrome.
    c. prognosis.
    d. diagnosis.

16. Any subjective phenomenon or departure from the normal function or structure or sensation experienced by the patient or client is a(n):
    a. prognosis.
    b. diagnosis.
    c. symptom.
    d. etiology.

17. When the structure and/or function of the human body is disabled by virus, bacteria, or tissue death, the study of this state is termed:
    a. proctology.
    b. parasitology.
    c. pathology.
    d. psychology.

18. In breast feeding, the harder and more frequently the infant suckles, the more milk is produced and secreted from the mammary glands and ducts. This phenomenon is called:
    a. metabolism.
    b. anabolism.
    c. negative feedback.
    d. positive feedback.

19. The medical term for the outer layer of skin is:
    a. dermatitis.
    b. epidermis.
    c. hypodermis.
    d. paradermotomy.

20. Blood pressure is measured in terms of:
    a. cubic centimeters.
    b. millimeters of mercury.
    c. liters.
    d. IbHg.

21. In the term *hypoglycemia*, its prefix is:
    a. hypo.
    b. glyc/o.
    c. emia.
    d. glycemia.

22. If *enter/o* means intestine, what is painful, inflamed intestine called?
    a. Enterology
    b. Paraenterotomy
    c. Enteritis
    d. Anenteromenorrhea

23. If *phleb/o* means vein, what is inflammation of a vein called?
    a. Phlebitis
    b. Microphlebosis
    c. Acroplebplasty
    d. Plebalgiosis

24. Which of the following is considered a vital sign?
    a. Skin color
    b. Stiffness
    c. Pulse
    d. Nausea

25. A physician who specializes in hearing disorders is an:
    a. audiology.
    b. audiologist.
    c. audiotomy.
    d. audiophobia.

26. A disease of unknown origin is:
    a. nosocomial.
    b. epidemiological.
    c. idiopathic.
    d. communicable.

27. A disease that can be spread from person to person is:
    a. contagious.
    b. communicable.
    c. nosocomial.
    d. epidemic.

28. A disease-producing organism is called:
    a. harmless.
    b. pathogenic.
    c. pathological.
    d. inflammatory.

29. The _____ response has the following signs and symptoms: redness, heat, swelling, and pain.
    a. inflammatory
    b. immune
    c. positive
    d. negative

30. An infection carried by an insect or animal is _____ borne.
    a. contact
    b. vector
    c. air
    d. water

31. Improperly sterilized medical equipment transmits infection via:
    a. contact transmission.
    b. common vehicle transmission.
    c. vector-borne transmission.
    d. airborne transmission.

32. _____ is spread by airborne transmission.
    a. Malaria
    b. Poison ivy
    c. HIV/AIDS
    d. Tuberculosis

33. When performing surgery, one should take the following precaution(s):
    a. gloves.
    b. gown.
    c. mask and eyewear.
    d. All of the above.

34. This is the single most important practice to reduce infection in health care settings.
    a. Masks
    b. Eyewear
    c. Hand washing
    d. Gown

35. Long-term health effects associated with a disease are known as:
    a. mortality.
    b. morbidity.
    c. signs.
    d. syndrome.

 # MATCHING EXERCISES

Set 1

Please match each term with the appropriate definition.

_____	1.	Phag/o	a.	Sugar
_____	2.	Leuk/o	b.	Blood
_____	3.	Hepat/o	c.	Vessel
_____	4.	Glyc/o	d.	Joint
_____	5.	Erythr/o	e.	Liver
_____	6.	Dermat/o	f.	Red
_____	7.	Angi/o	g.	Bone
_____	8.	Gastr/o	h.	Swallow
_____	9.	Oste/o	i.	Stomach
_____	10.	Arthr/o	j.	Skin
			k.	Intestine
			l.	White

Set 2

Please match each term with the appropriate definition.

_____	1.	Slow	a.	Otomy
_____	2.	Pain	b.	Hyper
_____	3.	Difficult	c.	Algia
_____	4.	Cutting into	d.	Peri
_____	5.	Within	e.	Hypo
_____	6.	Surgical removal of	f.	Tachy
_____	7.	Small	g.	Penia
_____	8.	Above normal	h.	An
_____	9.	Below normal	i.	Brady
_____	10.	Decrease or lack of	j.	Endo
			k.	Ectomy
			l.	Micro
			m.	Dys

Set 3

Please match each description with the appropriate term.

_____	1.	Death rate attributed to a disease
_____	2.	When the body breaks down substances into smaller components usable as energy sources
_____	3.	When the body uses raw material like amino acids to build certain structures as protein molecule chains
_____	4.	*Osteo/o* in the word *osteochondritis*
_____	5.	Long-term complications from a disease or condition
_____	6.	Client A feels tired and lethargic, her skin is flushed, and she has a rapid pulse. The attending health care professional believes she has water intoxication (hyponatremia), probably from the marathon run 2 days prior. In terms of the presented disorder/dysfunction, the lack of electrolytes during the marathon would be its
_____	7.	Client A feels tired and lethargic, her skin is flushed, and she has a rapid pulse. The attending health care professional believes she has water intoxication (hyponatremia) probably from the marathon 2 days prior. In terms of the client's condition, being tired and lethargic represents her
_____	8.	Client A feels tired and lethargic, her skin is flushed, and she has a rapid pulse. The attending health care professional believes she has water intoxication (hyponatremia), probably from the marathon run 2 days prior. In terms of the presented disorder/dysfunction, *hyponatremia* is the
_____	9.	Client A feels tired and lethargic, her skin is flushed, and she has a rapid pulse. The attending health care professional believes she has water intoxication (hyponatremia), probably from the marathon run 2 days prior. In terms of her condition, the rapid pulse would be its
_____	10.	The mechanism, also termed vicious cycle, by which the body continues the response or magnifies the response to a stimulus

a. Anabolism
b. Morbidity
c. Mortality
d. Acronym
e. Symptom
f. Syndrome
g. Etiology
h. Narcosis
i. Catabolism
j. Suffix
k. Sign
l. Positive feedback
m. Negative feedback
n. Prefix
o. Root word
p. Diagnosis

Set 4

Please match each term with the appropriate definition

_____	1.	Endemic	a.	Worldwide infection
_____	2.	Epidemic	b.	Cause of disease
_____	3.	Pandemic	c.	Spread by insect bite
_____	4.	Nosocomial	d.	Disease in large numbers in specific region
_____	5.	Airborne	e.	Spread to open wound by fluids
_____	6.	Vector-borne	f.	Cycle of infection
_____	7.	Direct contact	g.	Acquired in hospital
_____	8.	Chain of infection	h.	Spread by droplets
_____	9.	Terminal	i.	Fatal
_____	10.	Etiology	j.	Occurring in a particular area

FILL IN THE BLANK

Fill in the blanks to complete the following statements.

1. The medical abbreviation for immediately:

   _____

2. The medical abbreviation for nothing by mouth:

   _____

3. The adjustment made in the human body to maintain a stable internal environment by opposing the stimulus:

   _____

4. The system of measurement most widely used in medical professions:

   _____

5. Using the principles of medical terminology, what is surgical repair of a vessel? _____

6. Using the principles of medical terminology, what is the study of the skin? _____

7. Using the principles of medical terminology, what is inflammation of the liver? _____

8. *Cholecyst/o* means gallbladder. Using the principles of medical terminology, the removal of the gallbladder is termed:

   _____

9. The general term for the physiological process that maintains a stable internal environment:

   _____

10. The cause of or a reasonable explanation for the manifestation of a disease: _____

11. The process of disease identification:

    _____

12. The prediction of a disease's outcome:

    _____

13. Cytology and histology are examples of
_____ anatomy.

14. Pain and swelling of joints are
_____ of arthritis.

15. Blood pressure, body temperature, and respiratory rate are examples of
_____ signs.

16. An infection acquired in a medical facility is called
_____.

17. A disease that spreads worldwide is called a(n)
_____.

18. A patient with three of the following: hyperglycemia, hypertension, high triglycerides, high blood cholesterol, and obesity, probably has this disorder: _____

19. The deliberate raising of body temperature by your body to fight infection is known as _____.

20. A(n) _____ is a specific group of signs and symptoms related to a particular disease.

21. The _____ is the reason a patient is seeking medical help.

22. Hay fever is an example of a(n)
_____.

23. When consumable goods become contaminated and cause disease, this is called _____ transmission.

24. Germs are attacked by your
_____ response.

25. You need to take only this precaution when talking to a patient:
_____

SHORT ANSWER

1. Contrast the terms *sign* and *symptom*.

_____

_____

_____

_____

2. Explain the two subdivisions of metabolism.

_____

_____

_____

_____

3. What does the study of anatomy and physiology entail?

_____

_____

_____

_____

4. What countries use the mathematical system based on the British Imperial system?

_____

_____

_____

_____

5. How does the body maintain homeostasis in a very cold environment?

_____

_____

_____

_____

6. List and briefly describe the routes of transmission for communicable diseases.

_____

_____

_____

_____

7. What is the difference between communicable and contagious?

_____

_____

_____

_____

CASE STUDY

To celebrate her 40th birthday, Jen booked herself and five of her friends on a cruise to Mexico. They were at their most adventurous, eating, drinking, and buying whatever they wanted. It was the trip of a lifetime. They had a ball. However, a couple of weeks after they got home, they became quite ill, with nausea, diarrhea, and fever. They even started to turn yellow! They were diagnosed with a viral infection of the liver that causes inflammation of the organ. Though the disease sometimes can be quite serious, Jen and her friends recovered. When they asked the doctor how they got sick, he told them the virus was usually transmitted in contaminated food.

1. This virus causes liver inflammation. What is the medical term for the disease?

   _____

   _____

2. Given how this virus was transmitted in this case, is it contagious or communicable?

   _____

   _____

3. What is the route of transmission of the disease?

   _____

   _____

4. How could the group have protected themselves from the disease?

   _____

   _____

LEARNING ACTIVITIES

1. Using a medical dictionary, see if you can tell what a word means by breaking it up into prefix, word root/combining form, and suffix. Each student should pick one word for the others to figure out.

2. Pick a disease and research its signs and symptoms on the Internet. List them.

3. Play medical terminology "Operation." For as many surgeries as you can think of, try to determine the medical term for the surgery, using the roots, prefixes, and suffixes you already know. Use the Internet or a medical dictionary to confirm that your terms are correct.

4. Play medical terminology "Pictionary." One student draws a picture of a medical term while other students guess the correct term.

5. Review standard precautions by running medical scenarios. One student should come up with a scenario while other students decide what precautions should be taken in that scenario. Use the chart in the book to be sure the answers are correct.

THE HUMAN BODY: READING THE MAP

Chapter 2

CHAPTER SUMMARY

For medical professionals to talk about the body specifically, they need to use specific terminology to explain locations on the body. This chapter will introduce you to the specific anatomical terms for parts of the body. All terms are in respect to a position called anatomical position. A person in anatomical position is standing face front, feet pointed forward, with arms hanging at the side and palms forward. In the hospital, patients may be placed in one of several different positions. Patients may be prone (facedown), supine (faceup), in Fowler's position (sitting partly reclined) or in Trendelenburg position (head lower than feet). No matter the position of the patient, directions are relative to anatomical position. Most important, "left" and "right" refer to the patient's left and right, not yours! Position may affect patient health. For example, standing quickly after lying down may cause a brief decrease in blood pressure known as orthostatic hypotension. Some patients may experience orthopnea, which is difficulty breathing while lying down. Patients with heart failure may have distended neck veins when sitting.

To describe specific parts of the body, medical professionals often use directional terms instead of common terms. Terms like front and back are not specific enough, so anterior and posterior are used instead. Each direction on the body has a specific term. In addition, in illustrations, the body is often cut into sections that allow the organs to be visible. A transverse section divides the body into superior (top) and inferior (bottom) parts. A sagittal section divides the body into left and right parts. A frontal (coronal) section divides the body into anterior (ventral) and posterior (dorsal) parts.

Internally, the body has several cavities. Two dorsal cavities, the cranial and spinal cavities, house the brain and spinal cord, respectively. Two ventral cavities, the thoracic and the abdominopelvic cavities, are separated by the diaphragm, the chief breathing muscle. The thoracic (chest) cavity is further divided into the mediastinum, pericardial cavity, and pleural cavities. The mediastinum is the center of the thoracic cavity. The pericardial cavity contains the heart, and the pleural cavities contain the lungs. The abdominopelvic cavity, inferior to the diaphragm, is divided into the abdominal and pelvic cavities by an imaginary line at the top of the hips. The abdominal and pelvic cavities contain most of your organs. The abdominopelvic cavity can be subdivided into sections, either nine regions or four quadrants. Most often, the quadrant system is used to describe the location of pain or bruising. In addition, each body part has a medical term associated with it. These terms are more specific than common terms such as leg or arm. Medical professionals must be familiar with the medical terms.

To view the inside of the body for diagnostic purposes, a variety of imaging techniques are used. X-rays use radiation to produce a flat photonegative image of the internal features of the body. Computerized tomography (CT scans) use multiple finely focused x-rays to produce a three-dimensional image. Magnetic resonance imagery (MRI) uses magnetic energy to produce three-dimensional images, sometimes of better quality than CT scans. Ultrasound uses sound waves to take real-time images of the body, allowing visualization of moving parts.

CHAPTER OUTLINE

  I. The map of the human body
      A. Anatomical position
      B. Other positions
          1. Prone
          2. Supine
          3. Trendelenburg
          4. Fowler's
      C. Body planes
          1. Transverse
          2. Median
          3. Coronal
      D. Directional terms

 II. Body regions
      A. Body cavities
          1. Dorsal
              a. Cranial
              b. Spinal
          2. Ventral
              a. Thoracic
                  (1) Mediastinum
                  (2) Pleural cavities
                  (3) Pericardial cavity
              b. Abdominopelvic
                  (1) Abdominal
                  (2) Pelvic
      B. Abdominal regions
      C. Abdominal quadrants
      D. Body terminology

III. Radiology
      A. X-rays
      B. CT scan
      C. MRI
      D. Ultrasound

MEDICAL TERMINOLOGY REVIEW

Define the following terms.

1. Patellar: _____

2. Plantar: _____

3. Antecubital: _____

4. Femoral: _____

5. Axillary: _____

6. Carpal: _____

7. Cervical: _____

8. Lumbar: _____

9. Umbilical: _____

10. Thoracic: _____

MULTIPLE CHOICE

Circle the letter of the correct answer.

1. What structure separates the thoracic cavity from the abdominopelvic cavity?
   a. The navel
   b. The diaphragm
   c. The nipple
   d. The liver

2. A slice through the human body that parallels the long axis and extends from front to back, dividing the body into left and right halves, is called the:
   a. frontal plane.
   b. median plane.
   c. horizontal plane.
   d. midtransverse plane.

3. Which is *not* part of the dorsal cavity?
   a. The oral cavity
   b. The spinal canal
   c. The cranium
   d. All of the above

4. The appendix is found in which abdominal quadrant?
   a. Hypogastric
   b. Left inguinal
   c. Right lower
   d. Left lower

5. What is the correct term for the area anterior to the elbow, marked by the flex of the elbow, superficial veins, and a slight depression?
   a. Antecubital
   b. Anteradial
   c. Antebrachial
   d. Axillary

6. In anatomical position, how is the body positioned?
   a. Sitting with back straight, chest out, feet flat on the floor, and palms in neutral position
   b. Body erect, face and feet pointing anteriorly
   c. Palms facing anteriorly, arms at the side
   d. Both b and c

7. Which plane divides the body and its parts into superior and inferior portions?
   a. Sagittal
   b. Midsagittal
   c. Cranial
   d. Transverse

8. Nearest to the point of origin is called:
   a. distal.
   b. anterior.
   c. proximal.
   d. superior.

9. The axillary region can be used to take temperature. Where is it?
   a. Armpit
   b. Ear
   c. Rectum
   d. Belly button

10. Toward the head is:
    a. superior.
    b. dorsal.
    c. inferior.
    d. distal.

11. In reference to the antebrachium, where is the hand?
    a. Superior
    b. Distal
    c. Deep
    d. Proximal

12. Near the surface is called:
    a. lateral.
    b. dorsal.
    c. ventral.
    d. superficial.

13. Where are the kidneys?
    a. Right and left upper quadrants
    b. Right and left lower quadrants
    c. Pelvic cavity
    d. Hypogastric region

14. In reference to the nose, where is the mouth?
    a. Superior
    b. Lateral
    c. Medial
    d. Inferior

15. In reference to the skull, where is the brain?
    a. Superficial
    b. Deep
    c. Anterior
    d. Posterior

16. Brown fat can accumulate in various parts of the body, including behind the knees. What is the clinical name of this area?
    a. Peroneal region
    b. Plantar region
    c. Patellar region
    d. Popliteal region

17. What is the back of the head area called?
    a. Orbital region
    b. Buccal region
    c. Cervical region
    d. Occipital region

18. The dorsal cavities consist of which two cavities?
    a. Right and left pleural
    b. Right and left cerebral hemispheres
    c. Cranial and spinal
    d. Thoracic and pelvic

19. What is another term for *ventral*?
    a. Anterior
    b. Dorsal
    c. Posterior
    d. Cephalic

20. What is another term for *posterior*?
    a. Ventral
    b. Anterior
    c. Dorsal
    d. Caudal

21. The majority of the stomach is in what quadrant of the abdomen?
    a. Umbilical
    b. Left upper
    c. Right hypochondriac
    d. Lower thoracic

22. In reference to the shoulders, where is the head?
    a. Deep
    b. Proximal
    c. Superior
    d. Caudal

23. What structures are contained in the pleural cavities?
    a. Lungs
    b. Trachea
    c. Esophagus
    d. All of the above

24. The thoracic cavity contains the following organs:
    a. Lungs, heart, and stomach
    b. Brain, spinal cord, and eyes
    c. Heart, lungs, and esophagus
    d. Stomach, spleen, and lungs

25. The mediastinum is a subdivision of which cavity?
    a. Umbilical
    b. Epigastric
    c. Pleural
    d. Thoracic

26. A patient reports pain in the lower right abdominal quadrant. Which disorder would you rule out first?
    a. Hepatitis
    b. Gastritis
    c. Pancreatitis
    d. Appendicitis

27. Difficulty breathing when lying flat is called:
    a. orthopnea.
    b. orthostatic hypotension.
    c. JVD.
    d. heart failure.

28. Pain during the psoas test, in which the patient raises his or her right leg while the doctor presses down on it, is diagnostic for:
    a. hepatitis.
    b. gastritis.
    c. pancreatitis.
    d. appendicitis.

29. To view damage to the brain, which is the most appropriate imaging technique?
    a. Ultrasound
    b. X-ray
    c. MRI
    d. Radiograph

30. Which is the correct order from least dense to most dense on an x-ray?
    a. Air, fat, water, bone
    b. Bone, air, fat, water
    c. Fat, water, air, bone
    d. Bone, fat, air, water

31. A narrowly focused x-ray that circles the body forming a 3-D image is a(n):
    a. MRI.
    b. CT scan.
    c. x-ray.
    d. ultrasound.

32. A patient presents at the emergency department with a gunshot wound to her mediastinum. She is bleeding profusely. What major organ is probably damaged?
    a. Lung
    b. Heart
    c. Liver
    d. Stomach

33. The disorder carpal tunnel syndrome is due to inflammation of a nerve in this part of the body:
    a. wrist.
    b. neck.
    c. foot.
    d. lower back.

34. Many people who stand for long periods of time suffer pain due to plantar fasciitis. Where is their pain?
    a. Ankle
    b. Knee
    c. Sole of foot
    d. Hip

35. A patient recently diagnosed with breast cancer is scheduled for a lymph node biopsy to check for cancer spread into nearby lymph nodes. Which part of her body will be checked first?
    a. Abdominal
    b. Inguinal
    c. Lumbar
    d. Axillary

MATCHING EXERCISES

Set 1

Please match each term with the appropriate definition.

_____	1. Prone	a. Toward the surface
_____	2. Superior	b. Faceup
_____	3. Lateral	c. Toward the tail
_____	4. Superficial	d. To the front
_____	5. Proximal	e. Away from the point of origin
_____	6. Distal	f. Away from the body's surface
_____	7. Deep	g. Away from midline
_____	8. Medial	h. Facedown
_____	9. Inferior	i. Toward the point of origin
_____	10. Supine	j. Toward the head
		k. To the back
		l. Toward midline

Set 2

Please match body region with structures contained within it.

_____	1. Cranial	a. Lungs, heart, and esophagus
_____	2. Mediastinum	b. Spleen
_____	3. Abdominal	c. Urinary bladder (be specific)
_____	4. Thoracic	d. Pancreas
_____	5. Dorsal	e. Brain (be specific)
_____	6. Pelvic	f. Lungs (be specific)
_____	7. Umbilical	g. Brain and spinal cord
_____	8. Ventral	h. Liver
_____	9. Left hypochondriac	i. Lungs, liver, and uterus
_____	10. Pleural	j. Belly button
		k. Pubis
		l. Heart and esophagus

Set 3

Please match each body part with the appropriate technical term.

_____	1. Fingers	a. Femoral
_____	2. Forearm	b. Pedal
_____	3. Foot	c. Sternal
_____	4. Breastbone	d. Antebrachial
_____	5. Neck	e. Orbital
_____	6. Wrist	f. Digital
_____	7. Thigh	g. Cervical
_____	8. Lower back	h. Axillary
_____	9. Eye area	i. Lumbar
_____	10. Mouth	j. Carpal
		k. Oral

FILL IN THE BLANK

Fill in the blanks to complete the following statements.

1. The opposite of ventral is
   _____.

2. The common name for the buccal region is the
   _____.

3. The _____ test assesses for appendicitis by applying resistant force to a raised right leg.

4. In the thorax, the only cavities that are paired are called the
   _____ cavities.

5. The plane that divides the body into anterior and posterior sections is called the _____ plane.

6. The plane that divides the body into superior and inferior sections is called the _____ plane.

7. The reproductive organs are located in the
   _____ cavity.

8. Above the right inguinal region and below the right hypochondriac region is the _____ region.

9. Below the umbilical region is a region known as the
   _____ region.

10. The inguinal region is also called the
    _____ region.

11. When a person is lying face downward, that is said to be the
    _____ position.

12. Medial to both the right and left hypochondriac regions is the
    _____ region.

13. In reference to the antebrachium, the brachium is
    _____.

14. In reference to the pleural cavities, the mediastinum is
    _____.

15. If the uterus is the point of origin and the vagina extends away from it, in clinical terms, the vagina is
    _____ to the uterus.

16. If the intestine pokes through a weakness in the abdominal wall, we call this a(n) _____.

17. Orthostatic hypotension may occur when a person changes posture. BP drops due to the effect of _____.

18. In medicine, "left" and "right" refer to the
    _____ (whose?) "left" and "right."

19. A patient has an injury to his/her lumbar vertebrae. Where is the injury?
    _____

20. _____ is an imaging technique that uses sound waves to take real-time images of the body.

21. To see a broken bone, the doctor would take a(n)

_____.

22. To view the exact location of a tumor in 3-D,

_____ is the best

technique.

23. _____ is the highest density on an x-ray.

24. If a patient hits the back of his head, he may have injured his

_____ region.

25. A cracked knee cap is a fractured

_____.

SHORT ANSWER

1. Describe the clinical divisions or quadrants of the abdominal region.

_____

_____

_____

_____

2. Give three examples in which the supine position is advantageous.

_____

_____

_____

_____

3. Using anatomical terms, direct an anatomist to the calf muscles from the patella.

_____

_____

_____

_____

4. Explain in clinical and directional terms the structures of the lower extremity from the hips to the toes.

_____

_____

_____

_____

5. Describe the three major planes of section.

   _____

   _____

   _____

   _____

6. List and briefly define the types of imaging techniques.

   _____

   _____

   _____

   _____

7. List the advantages and disadvantages of various imaging techniques.

   _____

   _____

   _____

   _____

LABELING ACTIVITIES

1. Shade each cavity with a contrasting color and list a major structure or organ contained in this cavity beneath the corresponding label using Figure 2–8 in your textbook as a guide.

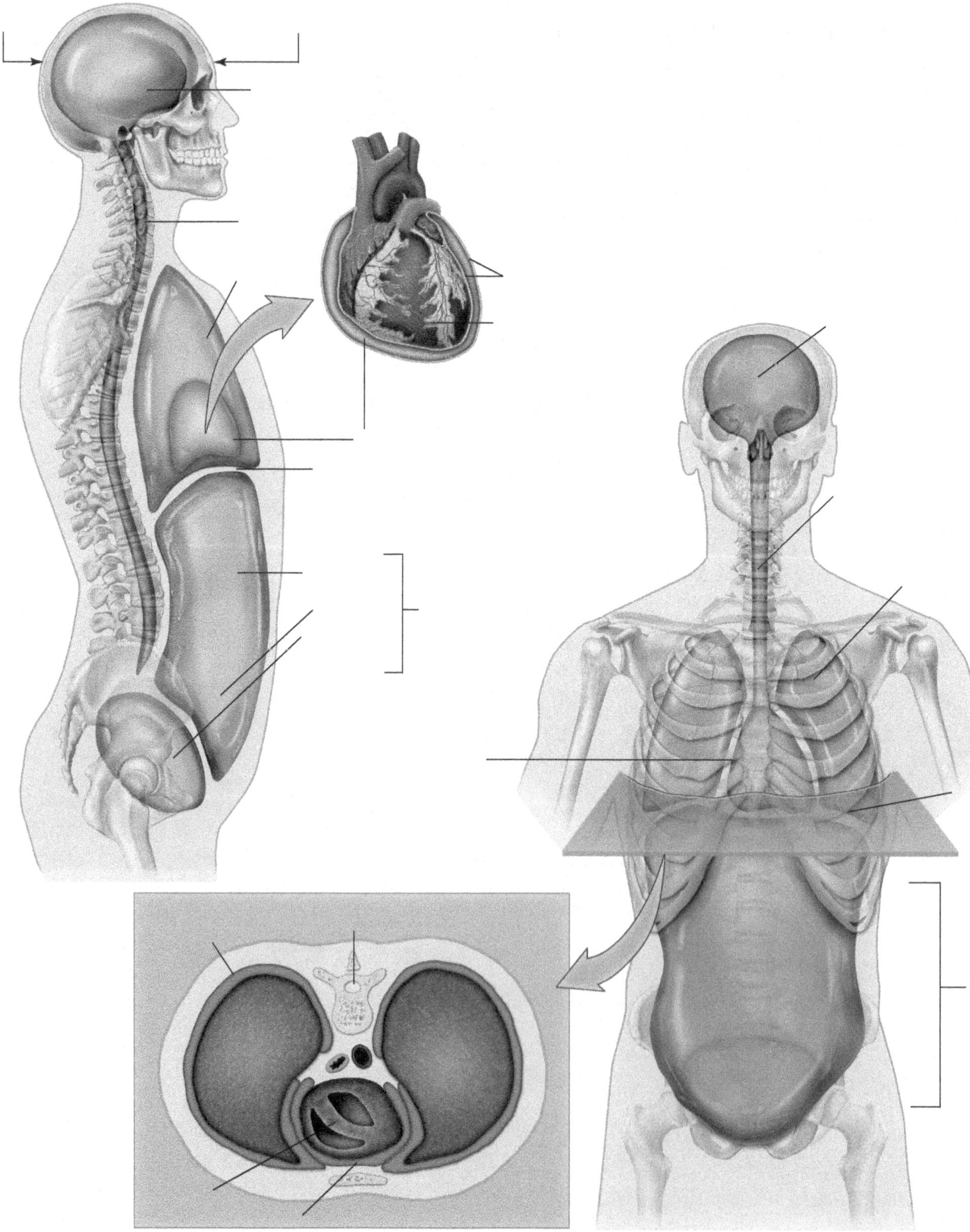

2. Label the regions of the body using Figure 2–13 in your textbook as a guide.

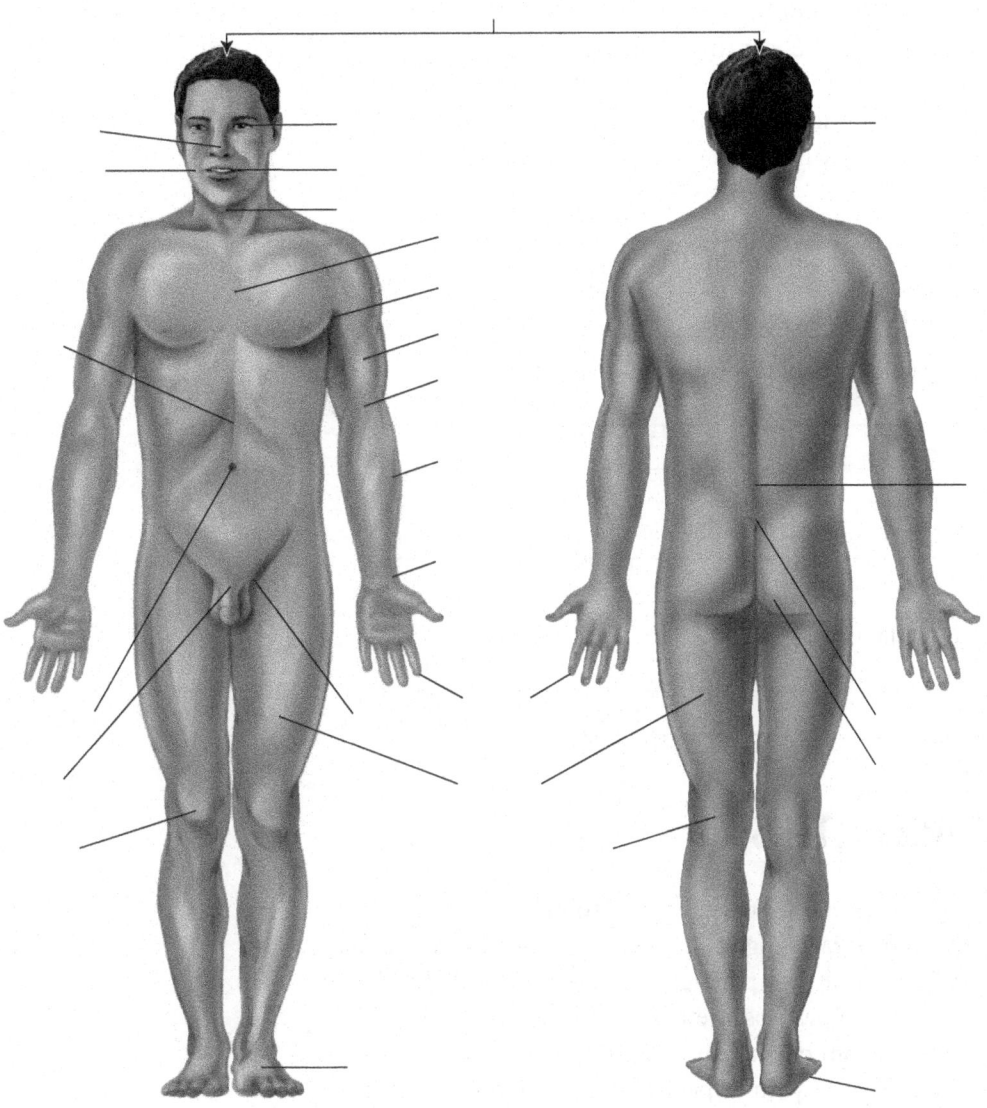

CASE STUDY

A 30-year-old male college professor slips on a piece of chalk and falls awkwardly, hitting his occipital region on the computer podium and landing face-first on the floor. When he regains consciousness at the hospital, he complains of carpal and patellar pain, as well as a headache. He has a large bruise on his occipital region and a cut that will need stitches above his left orbit. The emergency doctor sends him for x-rays and a CT scan. Using common terms, answer the following questions.

1. What will be x-rayed?

   _____

   _____

2. What part of the body will be scanned?

   _____

   _____

3. Where will he need stitches?

   _____

   _____

4. Why does he have a headache?

   _____

   _____

LEARNING ACTIVITIES

1. If blood pressure changes when posture changes, then pulse rate may also change. Speculate how pulse rate might change as posture changes, and then test your hypothesis. Split into partners and take the radial pulse of your partner as he or she changes posture from lying down to standing to sitting. Pool the results from other pairs of students. Does anything change? Does your data support your hypothesis? Do some research to find out how posture affects pulse rate.

2. Buy several different types of fruits and vegetables at the grocery store. Cut each into sections: sagittal, transverse, and frontal. How does each section differ?

3. Hide some "treasure." Write a treasure map using directional terms to guide other students to the treasure.

4. Pick a body part. Using the Internet, research the medical importance of the body part.

5. Play "Pin the Term on the Human." Make a cardboard cutout of a human shape. Pin the correct name on each part of the body.

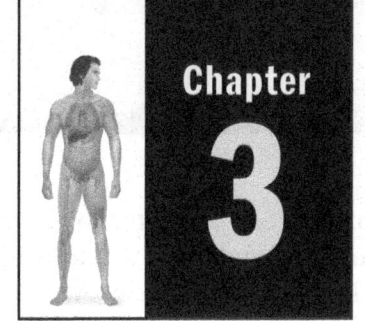

BIOCHEMISTRY: THE BASIC INGREDIENTS OF LIFE

CHAPTER SUMMARY

Your body is made of cells, and cells are made of molecules. There is no way to understand how your body works without understanding some basic biochemistry, the chemistry of living things.

All matter is made of elements, the smallest unit that retains the chemical nature of a substance. The smallest unit of an element is an atom. Atoms are made of protons, neutrons, and electrons. Hook together two or more elements, and you have a molecule. Many elements, called trace elements, are necessary for your cells to function. These elements are only needed in small quantities.

In atoms, protons are positively charged and electrons are negatively charged. Typically there are equal numbers of protons and electrons in an atom. If there are unequal numbers of protons and electrons, then that atom will carry a charge and will be considered polar. Charged atoms or molecules are called ions. Physiologically important ions are known as electrolytes. Acids and bases are special electrolytes that give up H^+ or OH^-. The acidity of a solution is measured on the pH scale.

In molecules, elements are held together by bonds between the electrons in the elements. If electrons are donated, the bond is ionic and the molecule is charged. If the electrons are shared, the bond is covalent and there is no charge. However, if the electrons are shared unequally, the molecule is polar and the bond is called polar covalent.

Water is the chief fluid in biological systems. It is polar because the bonds between hydrogen and oxygen are polar covalent. Because water is polar, only polar molecules will mix with water. Nonpolar will not. Molecules that will mix with water are called hydrophilic. Those that won't mix with water are called hydrophobic. The polarity of water also leads to hydrogen bonding. The hydrogen on one molecule binds weakly to the oxygen on another. This gives water unique properties including a high heat capacity, the ability to store heat.

Water is the chief biological fluid and thus is the basis for all solutions in the body. A solution is a solute dissolved in a solvent. In all biological solutions, the solvent is water. The amount of solute in a volume of solution is called the conentration.

All cells are made of molecules, and many of those molecules are biological molecules, carbon-based molecules found in living things. There are four major types of biological molecules: carbohydrates, lipids, proteins and nucleic acids. Each of the molecules has unique structures and chemical properties.

Cells are capable of making and breaking down biological molecules. Reactions that break down biological molecules are known as catabolism. Reactions that build molecules are known as anabolism. All the reactions are collectively called metabolism. Most of these reactions would happen very slowly in cells, too slowly to be useful to the cell. So cells have enzymes, molecules that speed up the rate of reactions. Enzymes are specific and are not used up in reactions.

Some of the most important reactions in cells are reactions that make energy. These reactions are collectively known as cellular respiration. Cellular respiration uses glucose and oxygen to make ATP, a high energy molecule used to power cellular reactions. Cellular respiration gives off carbon dioxide and water as waste products.

CHAPTER OUTLINE

I. Atoms, elements, ions
   A. Trace elements
   B. Atoms
   C. Parts of the atom
   D. Molecules
   E. Ions
   F. Electrolytes

II. Acids and bases

III. Bonding
   A. Ionic bonds
   B. Covalent bonds
   C. Polar covalent bonds

IV. Water
   A. Hydrophobic/hydrophilic
   B. Hydrogen bonds
   C. Solutions
      1. Solute
      2. Solvent
      3. Concentration

V. Biological molecules
   A. Carbohydrates
   B. Lipids
   C. Proteins
   D. Nucleic acids

VI. Metabolism
   A. Anabolism and catabolism
   B. Enzymes
   C. Cellular respiration and ATP

MEDICAL TERMINOLOGY REVIEW

Define the following terms.

1. Steroid: _____

2. pH: _____

3. Electrolytes: _____

4. Hydrophobic: _____

5. Solution: _____

6. Concentration: _____

7. Biological molecules: _____

8. Glucose: _____

9. Glycogen: _____

10. Metabolism: _____

MULTIPLE CHOICE

Circle the letter of the correct answer.

1. Cellular respiration makes ATP and uses:
    a. glycogen.
    b. glucose.
    c. water.
    d. carbon dioxide.

2. In an enzyme catalyzed reaction, the reactants are called:
    a. products.
    b. reactants.
    c. substrates.
    d. proteins.

3. Which of the following is true of enzyme reactions?
    a. Enzymes are not used up.
    b. Enzymes are nonspecific.
    c. Enzymes are carbohydrates.
    d. All of the above

4. Many catabolic reactions are _____ reactions.
    a. dehydration
    b. hydrophobic
    c. hydrolysis
    d. hyperactive

5. Which of the following is a function of proteins?
    a. Structure
    b. Nutrient storage
    c. Communication
    d. All of the above

6. Which of the following is a steroid?
    a. Cholesterol
    b. Ethanol
    c. Prevental
    d. Phospholipid

7. This is the most hydrophobic molecule known:
    a. phospholipid.
    b. cholesterol.
    c. triglyeride.
    d. wax.

8. Most carbohydrates have this function:
    a. enzyme.
    b. energy storage.
    c. structure.
    d. waterproofing.

9. Simple sugars have _____ carbons.
   a. four
   b. ten
   c. six
   d. eight

10. A solution is a _____ dissolved in a _____.
    a. solute; solution
    b. solute; solvent
    c. solvent; solute
    d. solution; solute

11. In _____ bonds, electrons are donated.
    a. covalent
    b. polar covalent
    c. ionic
    d. hydrogen

12. A solution with a pH of 8 is:
    a. acidic.
    b. neutral.
    c. basic.
    d. ionic.

13. In an atom, _____ surround the nucleus.
    a. electrons
    b. protons
    c. neutrons
    d. introns

14. These atomic particles are negatively charged:
    a. positrons.
    b. protons.
    c. neutrons.
    d. electrons.

15. This element is needed for formation of hemoglobin:
    a. sodium.
    b. manganese.
    c. calcium.
    d. iron.

16. Atoms that gain or lose electrons are called:
    a. elements.
    b. covalent.
    c. ions.
    d. acids.

17. Molecules that will mix with water are:
    a. hydrophobic.
    b. hydrophilic.
    c. hydrolysis.
    d. hydrogenated.

18. A molecule has several nitrogens in the backbone of the molecule. It is most likely a:
    a. carbohydrate.
    b. lipid.
    c. protein.
    d. nucleic acid.

19. A molecule has two hydrogens and one oxygen for every carbon. What kind of molecule is it?
    a. Carbohydrate
    b. Lipid
    c. Protein
    d. Nucleic acid

20. These molecules have both hydrophobic and hydrophilic portions:
    a. waxes.
    b. phospholipids.
    c. triglycerides.
    d. oils.

21. Ultimately we need to breathe because:
    a. we need oxygen to make proteins.
    b. we need oxygen to make nucleic acids.
    c. we need oxygen to make carbohydrates.
    d. we need oxygen to make ATP.

22. Which of the following is *not* a biological molecule?
    a. Ethanol
    b. Gasoline
    c. Glucose
    d. DNA

23. Egg white is this type of molecule:
    a. carbohydrate.
    b. lipid.
    c. protein.
    d. amino acid.

24. A molecule that is made of a phosphate, sugar, and base is a(n):
    a. amino acid.
    b. hydrochloric acid.
    c. nucleic acid.
    d. All of the above

25. Which of the following is a characteristic
    of enzymes?
    a. Saturation
    b. Specificity
    c. Competition
    d. All of the above

MATCHING EXERCISES

Set 1

Please match each term with the appropriate type of molecule.

_____	1.	Glucose	a.	Disaccharide
_____	2.	Sucrose	b.	Steroid
_____	3.	Egg-white albumin	c.	An electrolyte
_____	4.	DNA	d.	Monosaccharide
_____	5.	Cholesterol	e.	An acid
_____	6.	Glycogen	f.	Nucleic acid
_____	7.	Zn	g.	High energy molecule
_____	8.	Na+	h.	Polysaccharide
_____	9.	HCl	i.	Protein
_____	10.	ATP	j.	A trace element

Set 2

Please match each molecule with the appropriate function.

_____	1.	Zinc	a.	Enzyme systems
_____	2.	Copper	b.	Teeth and bones
_____	3.	Iron	c.	Thyroid gland
_____	4.	Manganese	d.	Cellular respiration
_____	5.	Iodine	e.	Hemoglobin
_____	6.	Fluorine	f.	Cell membrane
_____	7.	Oxygen	g.	CNS, fat metabolism
_____	8.	Water	h.	Genetic code
_____	9.	Phospholipids	i.	Chief biological solvent
_____	10.	DNA	j.	Amino acid metabolism

Set 3

Please match each molecule with the appropriate pieces.

_____ 1. Protein
_____ 2. DNA
_____ 3. Polysaccharide
_____ 4. Oil
_____ 5. Wax
_____ 6. Phospholipid
_____ 7. Molecule
_____ 8. Disaccharide
_____ 9. Carbohydrate
_____ 10. Enzyme

a. Glycerol and fatty acids
b. Monosaccharides
c. CH_2O
d. Amino acids
e. Atoms
f. Two monosaccharides
g. Fatty acid with alcohol
h. Protein with binding site
i. Nucleotides
j. Phosphate head and fatty acid tails

Set 4

Please match each term with the appropriate definition.

_____ 1. Metabolism
_____ 2. Saturation
_____ 3. Inhibition
_____ 4. Hydrolysis
_____ 5. Element
_____ 6. Hydrophilic
_____ 7. Dehydration synthesis
_____ 8. Concentration
_____ 9. Biological molecules
_____ 10. Steroids

a. Amount of solute dissolved in solvent
b. Enzyme is blocked
c. Can mix with water
d. All the chemical reactions in the body
e. Make molecules by removing water
f. Adding water to break down molecules
g. Molecules found in living things
h. Binding sites are full
i. Ringed lipids
j. Smallest unit of matter

FILL IN THE BLANK

Fill in the blanks to complete the following statements.

1. Two or more elements joined together form a(n)

   _____.

2. Elements are usually abbreviated using the

   _____ of their technical

   name.

3. A fluorine deficit may cause

   _____.

4. Positively charged particles found in the nucleus of an atom are

   _____.

5. Polar molecules are

_____.

6. Ions found in the body are

_____.

7. One of the main functions of the
_____ system is to regulate
electrolyte balance.

8. HCO_3^- is _____ ion.

9. A(n) _____ can release
hydrogen ions.

10. A bond with unequal sharing of electrons is a(n)
_____ bond.

11. In a solution, the substance doing the dissolving is the

_____.

12. Steroids are called "anabolic" because they

_____ tissues.

13. Enzymes _____ biological
reactions.

14. Only some substances can be carried by enzymes, thus enzymes are

_____.

15. ATP is made in this organelle:

_____.

16. The two waste products from cellular respiration are
_____ and

_____.

17. When ATP gives off energy, it loses a phosphate and becomes

_____.

18. Carbon dioxide is a weak

_____.

19. Most anabolic reactions are _____
reactions.

20. Negatively charged subatomic particles are

_____.

21. A special link called a(n) _____ is
found in proteins.

22. Water cools down more _____ than
air.

23. An atom is the smallest recognizable unit of a(n)

_____.

24. These molecules may be used for defense, communication, structure, and
muscle: _____.

25. Neutral pH is pH

_____.

SHORT ANSWER

1. Explain how enzymes work.

   _____

   _____

   _____

   _____

2. List the four classes of biological molecules and their characteristics.

   _____

   _____

   _____

   _____

3. Explain the three types of bonds.

   _____

   _____

   _____

   _____

4. Explain cellular respiration.

   _____

   _____

   _____

   _____

5. Explain metabolism.

   _____

   _____

   _____

   _____

LEARNING ACTIVITIES

1. Research how biochemistry relates to activities in a hospital laboratory.
2. Research various related professional careers in biochemistry.
3. Trace elements are very important for physiology. Using the Internet, find some diseases that involve either too much or too little of one of the trace elements. Discuss your findings. How many different disorders can you find as a group?

THE CELLS: THE RAW MATERIALS AND BUILDING BLOCKS

 CHAPTER SUMMARY

All living things are made of cells. Cells are the fundamental unit of living things, the building blocks of our bodies. Cells are surrounded by a semipermeable membrane that allows only some substances to pass into and out of the cell. Substances can be transported across the cell by either active or passive transport methods. Passive transport methods, including diffusion, osmosis, filtration, and facilitated diffusion, do not cost the cell anything. Substances transported by passive transport generally are moving down their concentration gradient from high to low concentration. Active transport methods, on the other hand, are expensive for cells because the cell must use ATP. Often substances transported actively are being transported against their concentration gradient from low to high concentration, like pumping water out of a flooded basement. Active transport methods include carrier mediated active transport, endocytosis, and exocytosis.

Cystic fibrosis, a genetic disease, occurs when ion channels in the cell membrane do not work well, causing the production of excess, sticky mucus. Diabetes mellitus, characterized by dangerously high blood sugar, is caused by the inability of glucose to get into cells. A hormone, insulin, is necessary for the passive transport of glucose into cells. Without insulin, or when cells are insensitive to insulin, glucose cannot get into cells, and it builds up in the bloodstream. Familial hypercholesterolemia is a genetic disorder caused by failure of endocytosis. In this disorder, cholesterol builds up in the bloodstream because it cannot be transported into cells. The buildup of cholesterol in the blood increases the risk of heart attacks and strokes.

Inside the cell is the cytoplasm: liquid and suspended organelles. Organelles are membrane-bound structures inside the cell with specific jobs. One could think of a cell like a small city, with each organelle fulfilling a specific duty necessary for the "city" to function. The nucleus is city hall, running the city. The genetic material, DNA, is included in the nucleus. Ribosomes are manufacturing plants that make proteins. The proteins are modified in the endoplasmic reticulum and packed and shipped by Golgi apparatus. Some of the proteins made by ribosomes are packed into lysosomes by Golgi apparatus. Lysosomes are the waste removal specialists, destroying molecules and bacteria. Mitochondria are the cell's power plants, making ATP via cellular respiration. When organelles malfunction, serious diseases result. Lysosomal storage diseases, like Tay-Sachs disease, are caused by missing lysosomal enzymes. Without these enzymes, molecules build up in cells, eventually destroying them.

Cell physiology is largely a matter of chemical reactions. Most chemical reactions that cells depend on are very slow, too slow to be of any use to a living cell. Thus, cells have enzymes, protein molecules that increase the rate of biological reactions. Without these enzymes functioning properly, cellular metabolism stops, and cell damage results. Phenylketonuria (PKU), a genetic disorder, is one example. Perhaps the most important example of chemical reactions necessary for cells is the group of reactions that is part of cellular respiration. Cellular respiration, which takes place in mitochondria, is the chief way that cells make adenosine triphosphate (ATP), the cell's energy molecule. Glucose and oxygen are necessary for most cells to make ATP. Carbon dioxide and water are waste products of the reaction. It is because of cellular respiration that it is so important that glucose gets into cells. Without glucose, cells use other materials, destroying body tissue, in a desperate attempt to make ATP. In patients with diabetes mellitus, some of the signs and symptoms of the disorder are caused by abnormal cell metabolism because glucose cannot get into cells.

Cells reproduce by dividing, making exact copies of themselves. The life of our cells can be described as a cell cycle divided into two phases: interphase, in which the cell is preparing to divide, and the mitotic phase, in which the cell is actually dividing. The mitotic phase is further divided into mitosis, the division and sorting of the genetic material from the cell's nucleus, and cytokinesis, the division of the cytoplasm. Because it is an absolute necessity that all the chromosomes (46 in humans) get copied and each new cell gets one of every chromosome, mitosis is divided up into several phases, prophase, metaphase, anaphase and telophase, based on the position of the chromosomes in the cell. Cell division is tightly controlled. Cells are usually prevented from dividing inappropriately. However, when the control system is damaged or broken, cells often will divide out of control. When cells divide out of control and spread to other parts of the body, cancer results. Cancer is life threatening because the spreading cells can invade distant organs and disrupt their function. Cancer is staged based on the amount of spread. Cancer diagnosed in more advanced stages is much harder to treat than cancer caught in early stages.

Microorganisms are single-celled or subcellular organisms. Microorganisms fall into four categories: bacteria, viruses, fungi, and protozoans. Bacteria are prokaryotic organisms with an amazing ability to reproduce. They are very simple cells with few organelles and no nucleus. Some bacteria are useful and necessary for human health, including a whole group of bacteria that live in our digestive system. Other bacteria are pathogens, invading our body and causing disease. Viruses are subcellular particles, consisting of a protein coat surrounding some genetic material, either DNA or RNA. Viruses can only reproduce by hijacking our cells and forcing the cells to produce millions of viruses. Viruses generally infect specific types of cells and are responsible for many common infections, including the common cold. Fungi, like our cells, are eukaryotic cells, with a nucleus and organelles. They spread via tiny filaments or by spore release. Most fungi are not pathogenic, although patients with compromised immune systems are often vulnerable to infection by fungi that are generally harmless. Mushrooms, for example, are fungi, as is the organism that causes athlete's foot. Protozoans are single-celled organisms with complex behavior. They are often found in water or transmitted by insects. Many are not pathogenic, but a few, like malaria, cause life-threatening illness.

Bacteria cause disease in many ways. They may cause direct tissue destruction, release toxins that destroy tissue or inhibit metabolism, induce blood clotting, or fill the lungs with fluid. Symptoms of bacterial infection include high fever, swelling, pain and discharge at infection site, and rapid pulse and respiration. Bacterial infection must be treated with antibiotics, which kill the bacteria without harming your cells. One of the most important issues in public health is the emergence of antibiotic-resistant bacteria. Because of overuse of antibiotics, many bacteria are no longer sensitive to standard antibiotics like penicillin. Often patients must use more than one antibiotic to get rid of bacteria that used to be easy to treat. Some bacteria are resistant to many antibiotics, making them especially dangerous.

Viral infections produce symptoms because cells are destroyed by the virus in its quest to make more of itself. Also, some viruses can actually put their genetic material into your cells, permanently altering their genetic makeup. Viral infections can sometimes become chronic or can reappear years after the first infection. Symptoms of viral infection include low-grade fever (though sometimes it is high), muscle aches, and fatigue. Some viruses may cause no symptoms at all. Viral infections may be dangerous because viruses often create an environment that encourages secondary bacterial infections. Viruses, because they are not cells, do not respond to antibiotics, and antiviral drugs typically harm the host cell. The most effective treatment for viral infection is rest, fluids, and treatment of symptoms. Most viral infections will be stopped by the immune system after several days.

Fungal infections are caused by inhalation of spores or the entrance of spores through a wound. Most fungi do not cause infection in healthy individuals, but a few cause skin infections, such as jock itch. Patients with immune systems that are suppressed may become very ill from fungal infection. Symptoms of fungal infection vary widely, and antifungal drugs are generally toxic because fungi are eukaryotic cells like our own cells.

Protozoan infection is generally caused by exposure to contaminated food or by transfer from a living vector like an insect. Symptoms, again, vary depending on the protozoan involved. Some diseases, like "beaver fever," are mild, but others, like malaria, are life threatening.

CHAPTER OUTLINE

I. Cell structure
   A. Cell membrane
      1. Structure
      2. Transport methods
         a. Passive
         b. Active
         c. Pathology
            (1) Diabetes mellitus
            (2) Cystic fibrosis
            (3) Familial hypercholesterolemia
   B. Cytoplasm
      1. Nucleus
      2. Organelles
      3. Pathology
         a. Organelle disorders
         b. Diabetes mellitus

II. Cell function
   A. Cellular respiration
   B. Enzymes
      1. Pathology: PKU
   C. Cellular reproduction
      1. Cell cycle
      2. Mitosis
      3. Pathology: cancer

III. Microorganisms
   A. Bacteria
   B. Viruses
   C. Fungi
   D. Protozoans
   E. Pathology
      1. Bacterial infections
      2. Viral infections
      3. Fungal infections
      4. Protozoan infections
      5. Antibiotic resistance

MEDICAL TERMINOLOGY REVIEW

Define the following terms.

1. Lysosomal storage disease: _____

2. Diabetes mellitus: _____

3. Cancer: _____

4. Metastasis: _____

5. Antibiotics: _____

6. Antibiotic resistance: _____

7. Familial hypercholesterolemia: _____

8. Pathogen: _____

9. Virus: _____

10. Bacteria: _____

 ## MULTIPLE CHOICE

Circle the letter of the correct answer.

1. The movement of water across a
    semipermeable membrane is called:
    a. diffusion.
    b. osmosis.
    c. phagocytosis.
    d. exocytosis.

2. Where is DNA synthesized?
    a. Endoplasmic reticulum.
    b. Lysosomes
    c. Golgi apparatus
    d. Nucleus

3. Which mechanism is active?
    a. Exocytosis
    b. Pinocytosis
    c. Endocytosis
    d. All of the above

4. The microorganism that causes herpes is a:
    a. bacterium.
    b. protozoan.
    c. virus.
    d. fungus.

5. A sperm cell propels itself with a single
    hairlike structure. This type of structure is
    called:
    a. flagellum.
    b. cilium.
    c. tinea cordae.
    d. cordae equina.

6. Containers A and B are separated by a
    semipermeable membrane. The solute
    concentration is 6 mg/mL in container
    A and 2 mg/mL in container B. In what
    direction will *osmosis* take place?
    a. B to A
    b. No movement can take place
    c. A to B
    d. From A to B first and then from B to A
        for stability

7. What process occurs across the walls of
    small blood vessels, pushing both water and
    dissolved nutrients into the tissues of the
    body?
    a. Osmosis
    b. Diffusion
    c. Filtration
    d. Hemolysis

8. What molecule stores instructions for
    protein synthesis?
    a. RNA
    b. DNA
    c. Protein
    d. Carbohydrate

9. When a membrane allows certain substances
    in and out, the membrane is said to be:
    a. semipermeable.
    b. selectively permeable.
    c. impermeable.
    d. Both a and b.

10. What must every cell have to maintain its integrity?
    a. Nucleus
    b. Cilia
    c. Cell membrane
    d. Capsid

11. The inner membrane of the trachea moves phlegm upward in a wavelike motion with its microscopic hairlike projections. These types of structures are called:
    a. flagella.
    b. cilia.
    c. endoplasmic reticula.
    d. receptors.

12. The type of cellular transport that moves substances against the concentration gradient is:
    a. diffusion.
    b. filtration.
    c. active pump.
    d. osmosis.

13. Glucose needs to be ushered into the cells using:
    a. facilitated diffusion.
    b. phagocytosis.
    c. pinocytosis.
    d. exocytosis.

14. How can viruses nourish themselves?
    a. They have a mouthlike opening that pulls in debris and cells floating in the blood or interstitial fluid.
    b. They must enter another cell and use that cell's parts for energy and growth material.
    c. As long as they are exposed to light, they need not nourish themselves.
    d. They nourish themselves through the absorption of methane outside cellular sources.

15. Which of the following is a type of movement across a cell membrane with or along the concentration gradient?
    a. Active transport pumps
    b. Diffusion
    c. Phagocytosis
    d. Exocytosis

16. The microorganism that causes athlete's foot is a:
    a. bacterium.
    b. virus.
    c. fungus.
    d. protozoa.

17. Postural muscles, such as muscles of the neck, are in constant need of energy. These muscle cells therefore contain and maintain higher quantities of certain organelles than do cells with smaller energy requirements. Which organelles?
    a. Mitochondria
    b. Nucleus
    c. Ribosomes
    d. Endoplasmic reticulum

18. What are the three main parts of a cell?
    a. Dendrite, axon, and soma
    b. Plasma membrane, cytoplasm, and nucleus
    c. Prophase, anaphase, and metaphase
    d. Cutaneous, serous, and mucous

19. A pathogen is:
    a. an organism that causes diseases.
    b. a host for viruses.
    c. a cellular receptor.
    d. an internal method of transport.

20. What is the function of the mycelia on fungi?
    a. Reproduction
    b. Absorb nutrients
    c. Cell division
    d. Movement

21. There are two classes of tumors. The life-threatening tumor is called:
    a. malignant.
    b. sigma.
    c. bacteria.
    d. benign.

22. When cardiac pressure forces plasma and various dissolved materials through the kidney membrane, this is an example of:
    a. diffusion.
    b. facilitated diffusion.
    c. filtration.
    d. reno-exocytosis.

23. An activated canister of tear gas is thrown into a room. Soon the gas has spread from wall to wall and floor to ceiling. This movement of the gas is an example of:
    a. diffusion.
    b. endocytosis.
    c. osmosis.
    d. pinocytosis.

24. When a cancerous tumor breaks off and travels to other parts of the body, it is said to be:
    a. thrombosized.
    b. embolized.
    c. metastasized.
    d. dormant.

25. Candidiasis is common in some people suffering from AIDS. Candidiasis is the result of what type of infection?
    a. Protozoan
    b. Fungal
    c. Viral
    d. Bacterial

26. Cystic fibrosis, a disorder characterized by excess mucus production, is caused by defective:
    a. mitochondria.
    b. ribosomes.
    c. membrane channels.
    d. lysosomes.

27. About 1 in 500 Americans have the moderate form of this disease, which causes heart attacks and strokes.
    a. Cystic fibrosis
    b. Diabetes mellitus
    c. Phenylketonuria
    d. Familial hypercholesterolemia

28. Which of the following is true of lysosomal storage diseases?
    a. They're genetic.
    b. They're due to missing or defective enzymes.
    c. They allow molecules to accumulate in cells.
    d. All of the above.

29. Type I diabetes is caused by:
    a. obesity.
    b. high blood sugar.
    c. immune attack on the pancreas.
    d. eating lots of candy.

30. Why must patients with phenylketonuria (PKU) avoid NutraSweet?
    a. It has too much sugar.
    b. It has too much sodium.
    c. It has too much phenylalanine.
    d. It has too much cholesterol.

31. A tumor that has spread to the lymph nodes and/or bloodstream is in stage:
    a. I.
    b. II.
    c. III.
    d. IV.

32. New cancer staging systems often take this into account:
    a. tumor extent and spread.
    b. tumor color.
    c. tumor age.
    d. tumor shape.

33. A patient presents at the Emergency Department with high fever, rapid pulse, and foul-smelling discharge and pain from a recent wound. What treatment would you recommend?
    a. Rest and fluids
    b. Antibiotics
    c. Antifungal medication
    d. Chemotherapy

34. These pathogens are prokaryotic cells:
    a. virus.
    b. fungi.
    c. protozoans.
    d. bacteria.

35. A child presents at the family doctor with the following symptoms: weight loss, excessive urination, fatigue, and excessive thirst. What test would you order?
    a. Sweat test
    b. Blood sugar
    c. Blood cholesterol
    d. Biopsy

MATCHING EXERCISES

Set 1

Please match each term with the appropriate definition.

_____ 1. Passive
_____ 2. Active
_____ 3. Diffusion
_____ 4. Filtration
_____ 5. Pinocytosis
_____ 6. Exocytosis
_____ 7. Endocytosis
_____ 8. Phagocytosis
_____ 9. Active transport pumps
_____ 10. Osmosis

a. Pressure is applied to force water and dissolved material across a membrane
b. Intake of liquid and food by cells by engulfing
c. Movement of substances from higher concentration to lower concentration
d. General term for a type of transport that requires energy
e. Movement of water from areas of low concentration of solute to areas with high concentration of solute
f. How a cell transports things out of itself using a vesicle
g. General term for a type of transport that requires no energy
h. "Pushing" more into the cell using ATP as energy
i. Specifically, the intake of liquid by cells by engulfing
j. Specifically, the intake of solid particles by cells by engulfing

Set 2

Please match each structure with its function.

_____ 1. Cell membrane
_____ 2. Nucleolus
_____ 3. Ribosome
_____ 4. Lysosome
_____ 5. Mitochondria
_____ 6. Endoplasmic reticulum
_____ 7. Golgi apparatus
_____ 8. Centrioles
_____ 9. Chromosome
_____ 10. Cytoplasm

a. A series of transport pathways in the cell, having two distinct forms
b. Where ribosomes are made
c. Produces ATP
d. Contains powerful enzymes
e. Made from DNA
f. Gel-like substance in which the cellular organelles float
g. Plays a critical role in cell division
h. Attaches to rough ER and produces protein
i. Surrounds the cells and allows certain substances in and other substances out
j. Packaging plant of a cell

Set 3

Please match each pathogen or disorder with the appropriate description.

_____ 1. Bacteria
_____ 2. Malaria
_____ 3. Thrush
_____ 4. Virus
_____ 5. Pathogenic
_____ 6. Capsid
_____ 7. Strep throat
_____ 8. Shingles
_____ 9. Fungus
_____ 10. Protozoa

a. Microorganism that contributes to the normal flora of the body; can be pathogenic or nonpathogenic

b. The coat that surrounds the genetic material of a virus

c. Reoccurrence of chicken pox

d. An adjective used when an organism is said to produce diseases

e. Microorganism that cannot reproduce or eat by itself; needs a host

f. A disease caused by a protozoan living inside mosquitoes

g. General term for one-celled, animal-like organisms responsible for many tropical diseases transmitted through consumption of unclean water

h. Organism that can be either one-celled or multicelled

i. A disease of the mouth caused by a fungal infection

j. A disease of the throat caused by bacterial infection

Set 4

Please match each pathogen or disorder with the appropriate treatment.

_____ 1. Type I diabetes
_____ 2. Type II diabetes
_____ 3. PKU
_____ 4. Cystic fibrosis
_____ 5. Tay-Sachs
_____ 6. Cancer
_____ 7. Strep throat
_____ 8. Chicken pox
_____ 9. Athlete's foot
_____ 10. Familial hypercholesterolemia

a. Diet low in phenylalanine
b. Antibiotics
c. Chemotherapy
d. Supplements, insulin, mucus thinners
e. Diet and exercise, medication
f. Rest, fluids, treat symptoms
g. Antifungal medication
h. Diet, cholesterol-lowering drugs
i. Fatal, no cure or treatment
j. Insulin injections

FILL IN THE BLANK

Fill in the blanks to complete the following statements.

1. The type of transport demonstrated by oxygen being transported from the lungs to the blood is

   _____.

2. When a cell surrounds a solid particle forming a vesicle and pulls it into the cells, this transport is called

   _____.

3. ATP stands for _____.

4. The situation in which more potassium is pulled into the cell despite being at a higher concentration inside the cell is called

   _____.

5. The blueprint of the cell is contained in genetic material called

   _____.

6. Of the two classes of tumors, the non-life-threatening one is a(n)
   _____ tumor.

7. The microorganism that is not killed by antibiotics is a(n)

   _____.

8. Certain bacteria in the intestine actually help synthesize vitamin

   _____.

9. Fungi can spread through the release of

   _____.

10. The substance that is dissolved in water is referred to as the

    _____.

11. The structural difference between ATP and ADP is the number of
    _____ groups.

12. The type of microorganism that makes up the normal flora of the human body is _____.

13. Most body cells possess a nucleus. The exception is the
    _____, which lacks a
    nucleus.

14. The smallest functional unit of the body is the

    _____.

15. A disease caused by a protozoan carried within the body of a mosquito is

    _____.

16. High blood sugar is the chief sign of

    _____.

17. The hormone _____ is
    missing or ineffective in patients with diabetes mellitus.

18. Familial hypercholesterolemia is caused by defective

    _____.

19. One of the reasons smoking is so bad for your health is that smoking paralyzes _____, often
    leading to chronic lung disorders.

20. Glucose must get into cells so cells can make this molecule, _____, a high energy molecule necessary for cell metabolism.

21. PKU, a genetic disorder, is caused by a missing/ineffective _____.

22. These organelles, _____, contain enzymes. If the enzymes are missing, serious genetic disorders result.

23. The removal of tissue for examination by a pathologist is called a(n) _____.

24. The treatment and course of any infection is determined by the _____ involved.

25. The best treatment for a(n) _____ infection is rest, fluids, and symptom management.

SHORT ANSWER

1. What is the function of the cell membrane?

   _____

   _____

   _____

   _____

2. Identify places where protozoa live.

   _____

   _____

   _____

   _____

3. What are the common features of rough and smooth endoplasmic reticula?

   _____

   _____

   _____

   _____

4. What is the difference between passive and active cellular membrane transport?

   _____

   _____

   _____

   _____

5. Describe the process of filtration.

    _____

    _____

    _____

    _____

6. Compare and contrast type I and type II diabetes mellitus.

    _____

    _____

    _____

    _____

7. Why don't antibiotics kill viruses? Why do antiviral drugs have so many side effects?

    _____

    _____

    _____

    _____

    _____

LABELING ACTIVITIES

1. Label the parts of the cell using Figure 4–11 from your textbook as a guide.

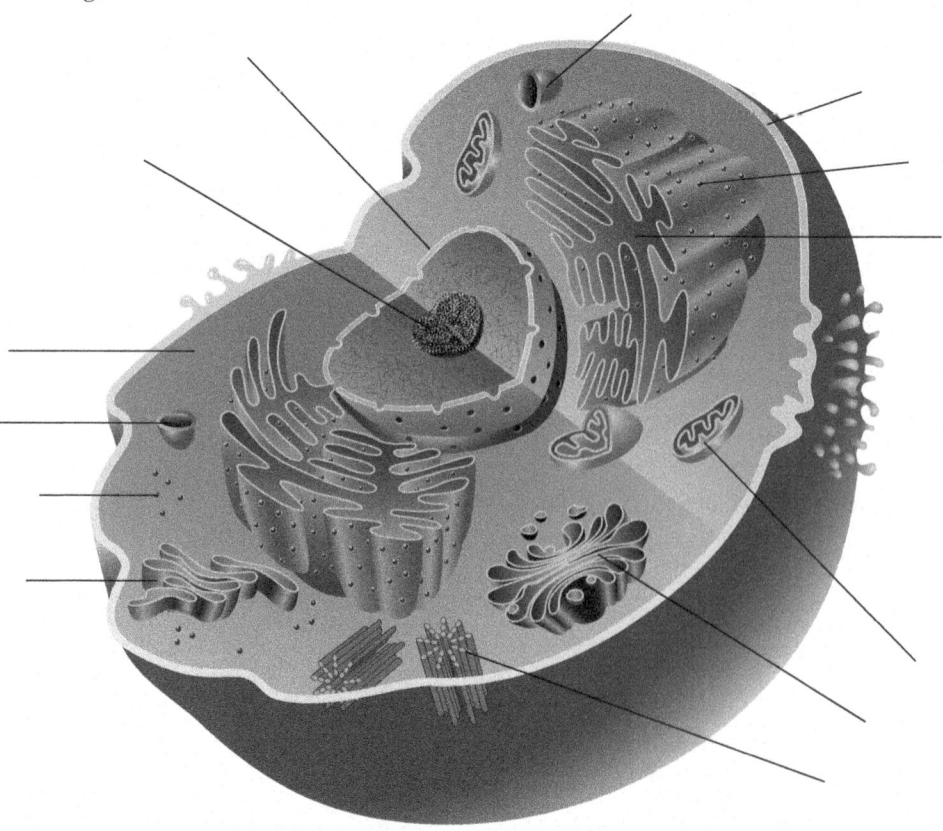

2. Label the parts of the cell membrane using Figure 4-3 from your textbook as a guide.

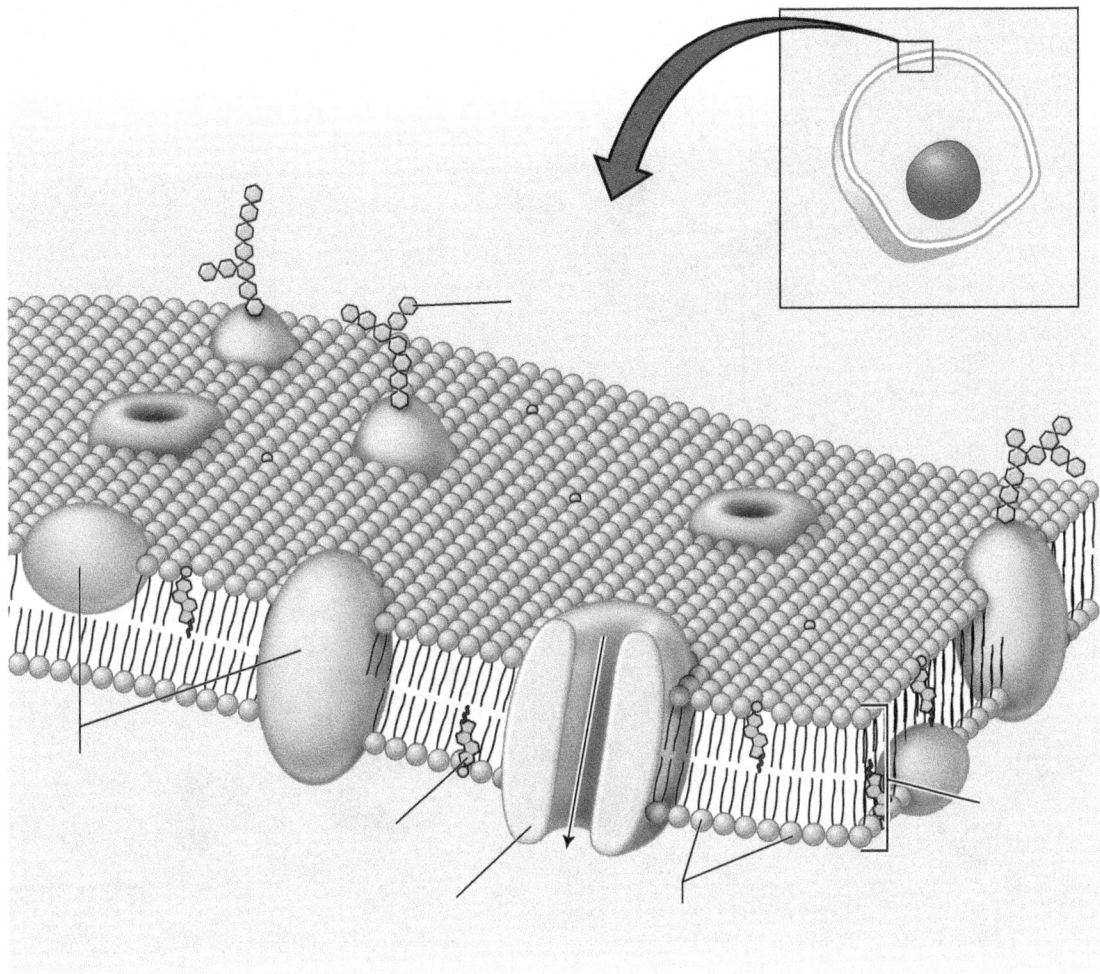

3. Label the parts of a bacterial cell and a virus using Figures 4–14 and 4–15
    from your textbook as a guide.

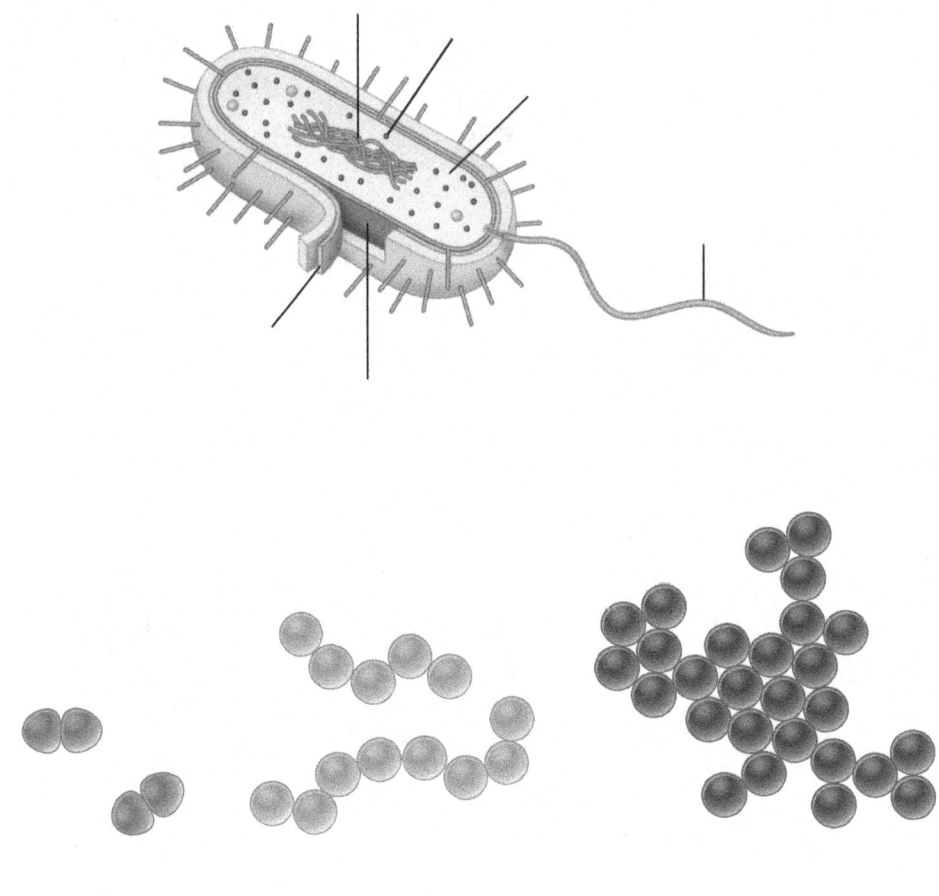

_____   _____   _____

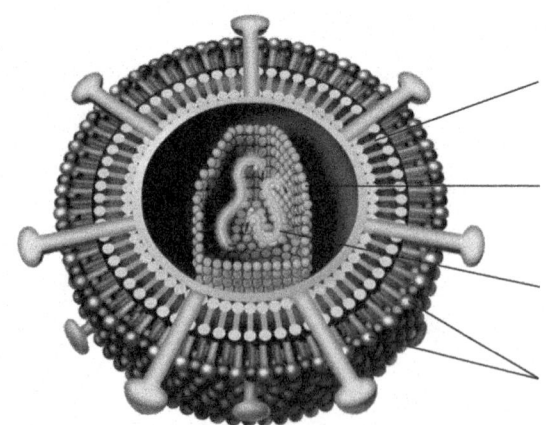

CASE STUDY

A four-year-old girl is brought to your office. She has been sick for two days. Her symptoms include a low-grade fever, headache, runny nose with clear discharge, fatigue, muscle aches, and a cough. She feels "yucky" but is clearly not bedridden. Her father is insisting you give her an antibiotic to cure her of the "flu."

1. Would you give her the antibiotic?

   _____

   _____

2. Why not?

   _____

   _____

3. Even if you don't think she needs an antibiotic, why not just give her the antibiotic to make the father happy? What's the harm in that?

   _____

   _____

LEARNING ACTIVITIES

1. Using the Internet as a source, list as many organelles and their disorders as you can.

2. For any type of cancer you choose, research the staging criteria. How are the criteria different from the general staging system in the textbook?

3. Play cell biology "Jeopardy." One person reads the definition of a term or the function of an organelle, and the other answers in the form of a question.

4. Design a pathogen board game. Use a pair of dice to determine how far pieces should move each turn. Make spaces like "Forgot to wash hands, go back 3 spaces" or "Caught MRSA, spend next turn in hospital." First one to make it to the end of the board, disease-free, wins.

5. Several Websites offer "virtual tours" of cells. Take a virtual tour. For each organelle, write down one new fact that you learned. Share your facts with other students in the class until you have generated a list of interesting facts for each organelle.

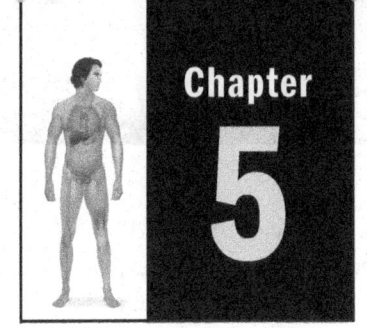

TISSUES AND SYSTEMS: THE INSIDE STORY

 ## CHAPTER SUMMARY

A collection of cells all working together is known as a tissue. Tissues can be divided into four basic types: epithelium, connective, muscle, and nervous. Epithelial tissue covers surfaces or lines cavities and is often specialized for secretion or absorption. Epithelium is classified based on the number of layers and the shape of the cells in the tissue. Some epithelial tissue is organized into membranes, which are sheetlike structures with specialized functions. Connective tissue is the most common tissue in the body. Connective tissue is a general support tissue and is found underlying all epithelial tissue. The most obvious characteristic of connective tissue is that the cells are embedded in a nonliving matrix. Connective tissue is classified based on the type of matrix. Muscle tissue allows the body to move. Skeletal muscle moves the skeleton, cardiac muscle moves the heart, and smooth muscle moves organs. Skeletal muscle is voluntary, whereas cardiac and smooth muscle are involuntary. Nervous tissue is a unique tissue confined to the nervous system. It consists of two types of cells, neurons, which send and receive information, and neuroglial cells, which support neurons.

Tissues may be damaged by pathogens and by chronic disease. Many pathogens may be prevented by the use of vaccinations. Many vaccines are universally recommended to prevent serious illness. Diabetes mellitus is often associated with circulatory problems. The decreased blood flow to tissues causes tissue damage and impaired wound healing. Melanoma is a form of skin cancer. Risk factors for melanoma include family history; excessive sun exposure; light hair, skin, or eyes; and having moles.

Different tissues are often grouped together into organs. Several organs that function together are an organ system. The skeletal system is made up of bones, and it supports, protects, and allows body movement. The muscular system is responsible for body movement. The integumentary system, or skin, is the covering of our body. It has several functions, including temperature regulation, touch sensation, and protection against infection and fluid loss. The nervous system, consisting of the brain, spinal cord, and associated nerves, is one of your control systems. It collects information, makes decisions, and formulates a response. The other control system is the endocrine system, a group of glands. Endocrine glands control cell function by releasing chemicals called hormones. The cardiovascular system, also known as the circulatory system, does just that. It circulates blood around the body. The heart pumps blood through a network of vessels that transport respiratory gases, nutrients, hormones, and waste products to every section of the body.

Many things can go wrong with the cardiovascular system. Clots can form, preventing blood from reaching body tissues, or pressure can build, rupturing the vessels. In addition, infection of the bloodstream, septicemia, can occur, allowing the pathogen to spread throughout the body. All these malfunctions are potentially fatal. The respiratory system is responsible for bringing oxygen into your body and taking carbon dioxide out. In addition, these gases diffuse between the blood and the lungs, allowing the gases to be transported around the body. The job of the lymphatic and immune system is to defend your body against infection. A series of general and specific mechanisms keep many pathogens from entering your body and defend you when the pathogen does get inside. The digestive (gastrointestinal) system is responsible for processing food and absorbing the nutrients contained in what you eat. In addition, what isn't used is removed from the body by this system. The urinary system, responsible for fluid and ion balance, filters blood and manufactures urine, removing waste from the body and reabsorbing water and chemicals. Last, the reproductive system is responsible for your ability to produce children. Male and female reproductive organs are very different, as are other physical characteristics controlled by hormones released by reproductive organs.

CHAPTER OUTLINE

I. Tissue types
   A. Epithelium
      1. Classification
      2. Membranes
   B. Connective
      1. Classification
      2. Membranes
   C. Muscle
   D. Nervous
   E. Pathology
      1. Vaccination
      2. Diabetes
      3. Melanoma

II. Organs

III. Organ systems
   A. Skeletal
   B. Muscular
   C. Integumentary
   D. Nervous
   E. Endocrine
   F. Cardiovascular
      1. Pathology: septicemia
   G. Respiratory
   H. Lymphatic and immune

I. Digestive
   1. Pathology: obesity
   2. Pathology: anorexia nervosa
   3. Pathology: bulimia
J. Urinary
K. Reproductive

MEDICAL TERMINOLOGY REVIEW

Define the following terms.

1. Meningitis: _____

2. Carrier: _____

3. Vaccination: _____

4. Atherosclerosis: _____

5. Melanoma: _____

6. Septicemia: _____

7. Multiple organ dysfunction syndrome: _____

8. Obesity: _____

9. Anorexia nervosa: _____

10. Bulimia: _____

MULTIPLE CHOICE

Circle the letter of the correct answer.

1. When an epithelial tissue is single layered and made of flat, scalelike cells, what is it called?
   a. Simple scalular
   b. Unisqualous
   c. Simple squamous
   d. Monoplatino

2. The most common type of tissue in the body is:
   a. epithelial.
   b. connective.
   c. nervous.
   d. muscular.

3. When an epithelial tissue is multiple layered and made of cells that are taller than they are wide, what are they called?
   a. Stratified columnar
   b. Striated towercity
   c. Polypillarity
   d. Multirectangular

4. What is the clinical term for *fat*?
   a. Synovial tissue
   b. Parietal tissue
   c. Adipose tissue
   d. Sebaceous tissue

5. What is the function of a neuron?
   a. Conductor of information
   b. Producer of hormone
   c. Protection and support
   d. Storage of fat and protein

6. Which of the following epithelial tissues can be found in the outermost layer of skin?
   a. Polypillarity
   b. Striated scalular
   c. Stratified squamous
   d. Simple columnar

7. Which of the following epithelial tissues can be found lining the air sac of the lungs?
   a. Striated scalular
   b. Simple squamous
   c. Monorectangular
   d. Unipillarity

8. Blood and lymph are considered to be:
   a. synovial tissue.
   b. connective tissue.
   c. serous tissue.
   d. mucous tissue.

9. The membrane that lines cavities that open to the exterior, such as the mouth, reproductive, and respiratory tracts, is called:
   a. mucous membrane.
   b. serous membrane.
   c. visceral membrane.
   d. cutaneous membrane.

10. The visceral layer of a serous membrane:
    a. lines the inside of the skull.
    b. wraps around the individual organs.
    c. holds neurons together.
    d. lines the body cavities.

11. Which of the following statements is true about the internal structure of skeletal muscle cells?
    a. Multinucleate
    b. The only organelles present in the cytoplasm are mitochondria, giving the cell a striped appearance
    c. Has no nucleus
    d. Has no organelles

12. The parietal layer of serous membranes:
    a. lines the body cavities.
    b. adheres to the brain and spinal cord.
    c. wraps around neuron to increase speed of impulse.
    d. lines the surface of the body.

13. A membrane called *mesentery* lines the abdominopelvic cavity and covers the organs. Taking this description into account, what type of membrane is this?
    a. Cutaneous
    b. Mucous
    c. Serous
    d. Pseudocuboidal

14. Because it involves conscious effort and thought for movement, skeletal muscles are also called:
    a. voluntary muscles.
    b. premeditated muscles.
    c. deliberate muscles.
    d. intentional muscle.

15. Which of the following correctly describes the function of the mucus that mucous membranes produce and secrete?
    a. Nourishment
    b. Drying agent
    c. Digestive enzyme
    d. Lubrication

16. Each organ is a group of several different kinds of:
    a. regions.
    b. systems.
    c. muscles.
    d. tissues.

17. Because it does not require us to consciously think about its contractions, cardiac and visceral muscle tissues are considered what kind of muscles?
    a. Habitual
    b. Spontaneous
    c. Involuntary
    d. Instinctive

18. Which of the organs below are paired?
    a. Spleens
    b. Kidneys
    c. Livers
    d. Urinary bladders

19. Body fat is what kind of tissue?
    a. Connective
    b. Epithelial
    c. Serous
    d. Cutaneous

20. Due to its microscopic appearance, visceral muscle is called:
    a. little Swiss.
    b. smooth.
    c. rough.
    d. spongy.

21. Which part of the neuron transmits impulses *toward* the cell body?
    a. Dendrite
    b. Soma
    c. Meninges
    d. Axon

22. The function of hormones is to:
    a. regulate metabolic processes.
    b. regulate fluid balance.
    c. regulate rate of growth and reproduction.
    d. All of the above

23. Where is visceral muscle found in the body?
    a. Digestive system
    b. Cardiovascular system
    c. Urinary system
    d. All of the above

24. Which of these is considered a vital organ?
    a. Gallbladder
    b. Appendix
    c. Brain
    d. All of the above

25. Which body system *stores* the mineral calcium?
    a. Circulatory
    b. Digestive
    c. Lymphatic
    d. Skeletal

26. The disease meningitis is an:
    a. inflammation of the uterus.
    b. inflammation of the membranes around the brain.
    c. inflammation of the mouth.
    d. inflammation of the finger.

27. Vaccines for tetanus are recommended for
    a. young children.
    b. unvaccinated teenagers.
    c. all adults.
    d. All of the above

28. A condition common in diabetics, in which blood circulation is impaired due to lipid deposits, is called:
    a. arteriosclerosis.
    b. atherosclerosis.
    c. multiple sclerosis.
    d. iron poor blood.

29. MODS is an abbreviation for:
    a. multiple organ dilation system.
    b. multiple organic dilapidation section.
    c. multiple organ dysfunction syndrome.
    d. multiple organ diaphoresis syndrome.

30. Rheumatoid arthritis is caused by immune attack on the synovial membranes in the body. What system is damaged in the disease?
    a. Muscular
    b. Nervous
    c. Immune
    d. Skeletal

31. Peritonitis, an inflammation and infection of the peritoneum, could be caused by which of the following conditions?
    a. Heart attack
    b. Ruptured appendix
    c. Broken arm
    d. Pneumonia

32. The inner lining of the digestive system is called a mucosa. What kind of membrane is it?
    a. Serous
    b. Mucous
    c. Cutaneous
    d. Meninges

33. If you have torn a ligament, what kind of connective tissue have you torn?
    a. Dense regular
    b. Adipose
    c. Areolar
    d. Blood

34. Melanoma is deadly because it can spread to distant organs via the lymph nodes. To which system do the lymph nodes belong?
    a. Cardiovascular
    b. Integumentary
    c. Skeletal
    d. Immune

35. Progressive severe weight loss in patients who deny they are losing weight is called:
    a. bulimia.
    b. anorexia nervosa.
    c. obesity.
    d. yo-yo dieting.

 ## MATCHING EXERCISES

Set 1

Please match each term with the appropriate description.

_____ 1. Nervous tissue
_____ 2. Muscle tissue
_____ 3. Connective tissue
_____ 4. Epithelial tissue
_____ 5. Striated
_____ 6. Stratified
_____ 7. Serous
_____ 8. Meninges
_____ 9. Cutaneous
_____ 10. Synovial membrane

a. Specifically, membranes that cover the spinal cord and brain
b. Tissue that holds things together and provides structure
c. Appearance of skeletal muscles
d. Is able to shorten itself; provides movement by and in our bodies
e. Epithelial tissue of more than one layer of cells
f. Found in space between joints and produces a slippery fluid
g. Membrane that covers organs and lines cavities
h. Communication; rapid messenger of information
i. Commonly known as the skin
j. The tissue type that covers the body and its parts

Set 2

Please match each system with its function.

_____	1. Lymphatic system
_____	2. Endocrine system
_____	3. Nervous system
_____	4. Female reproductive system
_____	5. Male reproductive system
_____	6. Digestive system
_____	7. Respiratory system
_____	8. Integumentary system
_____	9. Cardiovascular system
_____	10. Muscular system

a. Supports and sustains structure; framework of body

b. Movement; controls the diameter of blood vessels

c. Surface protection from harmful environmental invaders

d. Produces ova and houses the growing fetus

e. Communication and control through the release of hormones

f. Produces red blood cells

g. Chemically and mechanically breaks down food for use by the body

h. Communication; transmission of impulses

i. Transports water, oxygen, and nutrients to and from the cells of the body

j. Produces sperm

k. Maintains proper fluid balance in the body; helps fight disease

l. Supplies fresh oxygen for the blood to absorb

Set 3

Please match each organ with its system

_____	1. Both pancreas and testes
_____	2. Both vas deferens and penis
_____	3. Tonsils and lymph vessels
_____	4. Heart and veins
_____	5. Small intestine and gallbladder
_____	6. Kidneys
_____	7. Sweat and oil glands
_____	8. Brain and spinal cord
_____	9. Trachea and bronchi
_____	10. Uterus and fallopian tubes

a. The system that produces sperm

b. The system that produces urine

c. The system that digests and eliminates food

d. The system containing sensory and motor neurons

e. The system that produces hormones

f. The system that allows us to flex and extend bones at moveable joints

g. The system that covers and protects the body

h. The system that allows for the union of sperm and ova

i. The system that eliminates carbon dioxide from the body

j. The system that includes blood

k. The system that cleans up excess fluid and fights infections

Set 4

Please match each disorder with the appropriate system.

_____ 1. Arthritis
_____ 2. Heart attack   a. Digestive
_____ 3. Appendicitis   b. Reproductive
_____ 4. Common cold   c. Integumentary
_____ 5. Brain tumor   d. Lymphatic/immune
_____ 6. Poison ivy   e. Respiratory
_____ 7. Ruptured spleen   f. Cardiovascular
_____ 8. Kidney failure   g. Sensory
_____ 9. Deafness   h. Urinary
_____ 10. Prostate cancer   i. Skeletal
                                       j. Nervous

FILL IN THE BLANK

Fill in the blanks to complete the following statements.

1. Both visceral and parietal membranes are part of
   _____ membranes.

2. The organ with which cardiac muscle is associated is called the
   _____.

3. A neuron is one type of nerve cell; the other type of nerve cell is called
   _____.

4. The testes belong to the reproductive and
   _____ systems.

5. The pancreas belongs to the endocrine and
   _____ systems.

6. Hormones circulate through the
   _____ system.

7. Sight, hearing, touch, taste, and smell belong to the
   _____ system.

8. The skin has the ability to produce vitamin
   _____.

9. The part of the neuron called the
   _____ transmits impulses
   away from the cell body.

10. Pseudostratified _____
    tissues line the lower part of the digestive tract.

11. The common feature of the lungs and kidneys is that both are
    _____, so if one is
    damaged, you can still survive.

12. When muscles are striped in appearance, they are said to be
    _____.

13. The membrane that lines joints produces a fluid that reduces friction
    called _____.

14. Tendons and ligaments are composed of dense
_____ tissue.

15. The spleen belongs to the _____
system.

16. A person infected with a disease but who may not get sick is called a(n)
_____.

17. An injection to train your immune system to recognize a pathogen so you
don't get sick is called a(n)
_____.

18. Diabetes may be related to tissue damage because the disease inhibits
_____.

19. _____ is a skin cancer,
often starting in an existing mole, that may be rapidly fatal.

20. _____ is the common term
for septicemia.

21. An individual with _____ goes
on eating binges and then purges what he/she has eaten.

22. Patients who are obese have excess
_____ tissue.

23. _____ is the chief cause of
melanoma.

24. If you are not making enough red blood cells, what tissue may be
malfunctioning?
_____

25. If your pleura is inflamed, where is the pain?
_____

SHORT ANSWER

1. List and describe the four main types of tissues.
_____
_____
_____
_____

2. Differentiate among three main types of muscle tissue.
_____
_____
_____
_____

3. Describe the main components of nervous tissue.

   _____

   _____

   _____

   _____

4. Describe the functions of a serous membrane.

   _____

   _____

   _____

   _____

5. Discuss the similarities of the endocrine and the nervous systems.

   _____

   _____

   _____

   _____

6. Explain how diabetes mellitus increases the risk of tissue damage.

   _____

   _____

   _____

   _____

7. List the risk factors for melanoma.

   _____

   _____

   _____

   _____

LABELING ACTIVITY

Label each type of tissue using Figure 5-1 from your textbook as a guide.

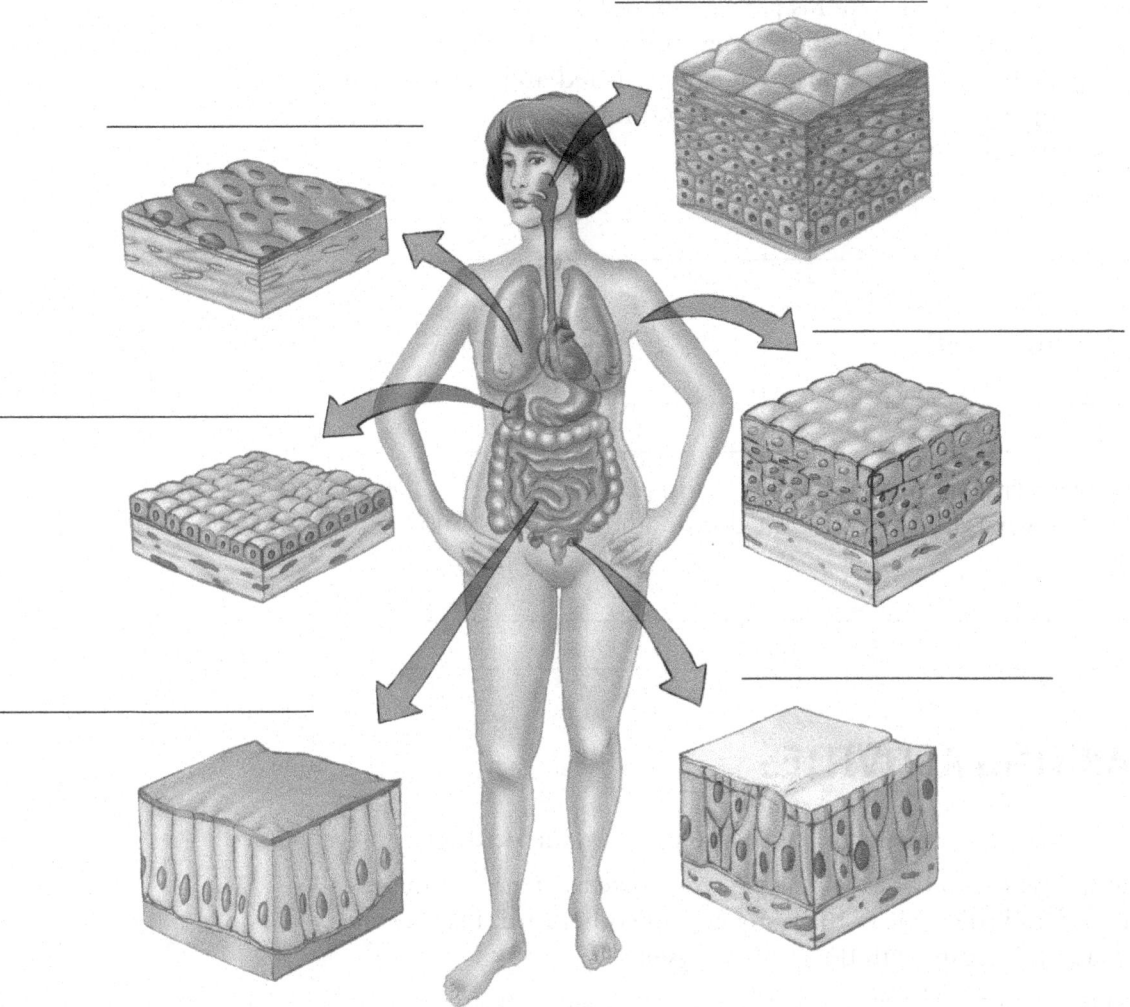

CASE STUDY

A 10-year-old boy playing near an abandoned mine falls and cuts his finger. Worried that he will get in trouble for playing in the mine area, he rinses his finger in the nearby stream and doesn't tell his parents about the injury. Within a few days, his finger is killing him, and he starts to feel really sick. He tells his parents, and they are appalled when they look at his hand. It is red and swollen. There is a red line running from his finger up his arm, and he has a fever. His parents rush him to the emergency department.

1. What is the diagnosis of his condition?

   _____

   _____

2. What is the treatment?

   _____

   _____

3. The emergency physician admits the boy and puts him on an IV, keeping him for several days. Why did the doctor admit the boy? What is the danger?

   _____

   _____

LEARNING ACTIVITIES

1. For each system, list as many organs as you can without looking.
2. Make a deck of cards. For each pair of cards, write the name of an organ on one card and the name of the system on the other card. Play "Go Fish," matching organs with their correct system.
3. Play tissue "Pictionary." One student can draw a tissue. Other students should guess the type of tissue based on cell shape and other physical characteristics.
4. Write a decision tree (dichotomous key) for the tissues to practice telling them apart. For each characteristic, answer yes or no for the tissue you are describing. Each decision will eliminate one type of tissue until you have arrived at the right identification by process of elimination. If you search for "dichotomous key to tissues," you will find several examples on the Internet. Once you have written the key, try using it to identify slides.
5. There are many different tissue disorders not covered in the book. Use the Internet to research disorders for one particular type of tissue.

THE SKELETAL SYSTEM: THE FRAMEWORK

CHAPTER SUMMARY

The skeleton is a living framework for your body, allowing movement, protecting delicate organs, and storing minerals. The skeleton consists of bones, united by joints. The bones themselves are classified by shape as either long, short, flat, or irregular. Each of the 206 bones in your body has a specific name that allows medical professionals to identify the bone that is injured. The outside of a bone is covered by connective tissue called the periosteum. Long bones are long and skinny. They have widened ends called epiphyses and a long shaft called the diaphysis. The surface of each epiphysis is covered by articular cartilage, which protects the bone at the joint. The bone surface is not smooth; it has many bumps, ridges, and hollows, which allow for muscle and ligament attachment and for joint surfaces.

Inside, long bones have a medullary (marrow) cavity and two different types of bones: spongy and compact. Spongy bone looks like a cleaning sponge, with holes between strips of bone called trabeculae. The trabeculae are covered by endosteum. Compact bone looks solid but isn't. It is composed of many tree trunklike structures called osteons. Osteons have a central canal for blood vessels and circular layers of bone surrounding the central canal.

Bone growth is called osteogenesis, or ossification. Several cells are involved in osteogenesis, including osteoprogenitor cells (which are bone stem cells), osteoblasts (which make bone), osteoclasts (which destroy bone), and osteocytes (which maintain the matrix). Bone development is of two types: intramembranous ossification, in which a connective tissue membrane turns to bone, and endochondral ossification, in which a hyaline cartilage model of a bone turns to bone. The bones of your skull are formed by intramembranous ossification. Most of the rest of your bones are formed by endochondral ossification. Even after your skeleton is fully formed, it is constantly being remodeled, broken down, and rebuilt by osteoclasts and osteoblasts. If breakdown is faster than buildup, osteoporosis may result. If a bone is fractured, it will be healed via endochondral ossification.

Joints are places where bones meet. At joints there are two other skeletal tissues: cartilage and ligaments (dense regular connective tissue). Joints may be classified by structure or by function (amount of movement), but the structure and function are usually related to each other. Most of the joints you can think of off the top of your head are synovial joints, with a large range of movement and a fluid-filled cavity between the bones. There are many different types of synovial joints and many different types of movement possible. Joint disorders, including several different types of arthritis, sprains, bursitis, and tendonitis are often painful and debilitating.

Skeletal injuries include disk herniation, abnormal spinal curvatures, and bone fractures of several different types. Joint and bone pathologies can be avoided by caring for your skeleton. Getting a proper diet and exercise, not smoking, and staying away from excessive caffeine can help protect your skeleton from the many things that can go wrong.

CHAPTER OUTLINE

I. Bone overview
   A. Shape
   B. External anatomy
      1. Diaphysis
      2. Epiphysis
      3. Periosteum
      4. Surface structures
   C. Internal anatomy
      1. Medullary cavity
      2. Spongy bone
      3. Compact bone
      4. Endosteum
   D. Bone tissue
      1. Osteons
      2. Trabeculae

II. Bone growth and repair
   A. Osteogenesis (ossification)
      1. Intramembranous
      2. Endochondral
   B. Pathology: osteoporosis

III. Other skeletal tissues
   A. Cartilage
   B. Ligaments and tendons

IV. Joints
   A. Structure
   B. Function/movement classification
   C. Pathology
      1. Arthritis
      2. Bursitis/tendonitis
      3. Sprain
      4. Dislocation/muscle tear
      5. Arthroscopic surgery

V. The skeleton
   A. Skull
   B. The bony thorax
   C. The spinal column
   D. Extremities
   E. Pathology
      1. Spinal pathology
      2. Bone fractures

VI. Bone and joint disorders

MEDICAL TERMINOLOGY REVIEW

Define the following terms.

1. Osteoporosis: _____

2. Arthritis: _____

3. Osteoarthritis: _____

4. Rheumatoid arthritis: _____

5. Tendonitis: _____

6. Sprain: _____

7. Arthroscopy: _____

8. Reduction: _____

9. Traction: _____

10. NSAID: _____

MULTIPLE CHOICE

Circle the letter of the correct answer.

1. When a fracture breaks the skin, it is known as a:
   a. simple fracture.
   b. closed fracture.
   c. compound fracture.
   d. Both a and b

2. Which of these is a nutritional disorder?
   a. Gigantism
   b. Warts
   c. Rickets
   d. Osteosarcoma

3. Osteomyelitis is an example of:
   a. an infection.
   b. a congenital disorder.
   c. a trauma.
   d. a tumor.

4. Which of the following allows your body to absorb ingested calcium from the digestive tract?
   a. Iron
   b. Vitamin B_{12}
   c. Vitamin D
   d. Phosphorus

5. Which of the following vices, according to your text, decreases bone mass?
   a. Caffeine: coffee or sodas
   b. Tobacco: cigarette smoking
   c. Bourbon: overindulgence in the spirits
   d. Chocolate: constantly feeding a sweet craving
   e. Both a and b
   f. a, b, c, d

6. What happens to ligaments and tendons as we age?
   a. They change from a bluish tint to an opaque yellow.
   b. They degenerate, becoming more flexible and loose, leading to increased range of motion, increased sprains, and more propensity for dislocations.
   c. They become less flexible, leading to decreased range of motion.
   d. They slowly detach from their attachment bones.

7. What is the function of osteoblasts?
   a. Tear down bone
   b. Build new bone
   c. Absorb calcium from the gut
   d. Stimulate calcium retention in the kidneys

8. Where are osteoprogenitor cells found?
   a. Periosteum
   b. Spleen
   c. Bone marrow
   d. Thymus

9. The primary component of the skeleton is:
   a. synovial fluid.
   b. cartilage.
   c. bone.
   d. ligament.

10. The phalanges and ulna are examples of what type of bone?
    a. Irregular
    b. Long
    c. Short
    d. Flat

11. The mandible and cervical vertebrae are examples of what type of bone?
    a. Irregular
    b. Long
    c. Short
    d. Flat

12. The parietal bones and scapulae are examples of what type of bone?
    a. Irregular
    b. Long
    c. Short
    d. Flat

13. The expanded ends of long bone are called:
    a. epimysia.
    b. epicondyles.
    c. epiphyses.
    d. epiosteum.

14. What substance is housed in the medullary cavity, yet absent in the trabeculae?
    a. Progenitor cells
    b. Yellow bone marrow
    c. Red bone marrow
    d. Digestive enzymes

15. What type of bony tissue makes up the adult diaphysis?
    a. Cancellous bone
    b. Spongy bone
    c. Cartilage
    d. Compact bone

16. Mature bone cells are clinically called:
    a. osteocytes.
    b. osteoblasts.
    c. osteoprogenitor cells.
    d. osteoclasts.

17. The formation of red blood cells is called:
    a. hemopoiesis.
    b. hemoglobin.
    c. hematocytosis.
    d. erythrogenesis.

18. Nodding the head in an aggressive gesture of "yes" is employing:
    a. adduction/abduction.
    b. rotation.
    c. flexion/extension.
    d. supination/pronation.

19. Connective tissue that attaches muscle to bone is called:
    a. tendon.
    b. cartilage.
    c. ligament.
    d. fascia.

20. Moving the joints of the ankle and foot so that the sole of one foot is facing away from the other is an example of:
    a. pronation.
    b. eversion.
    c. hyperadduction.
    d. plantar flexion.

21. Which of the following bones belongs to the axial skeleton?
    a. Clavicle
    b. Scapula
    c. Hyoid
    d. Tarsal

22. Which of the following bones belongs to the appendicular skeleton?
    a. Ribs
    b. Sternum
    c. Ilium
    d. Sacrum

23. The tip of the sternum is called the:
    a. xyphoid.
    b. hyoid.
    c. condyloid.
    d. patella.

24. The vertebral column has how many vertebrae in the midbuttocks region, neck region, lower back, and upper back region, respectively?
    a. 7, 12, 5, 5
    b. 3-4, 12, 5, 5
    c. 5, 7, 5, 12
    d. 1-4, 7, 12, 5

25. Besides depression and elevation, which of the following are also actions of the human mandible?
    a. Protraction and retraction
    b. Supination and pronation
    c. Inversion and eversion
    d. Flexion and extension

26. Which of the following is *not* true of osteoporosis?
    a. Risk increases with age.
    b. Caucasians and Asians are at higher risk.
    c. Men cannot get it.
    d. Smoking and sedentary lifestyle increase risk.

27. _____ is a type of arthritis due to buildup of uric acid in blood. It is affected by diet.
    a. Osteoarthritis
    b. Rheumatoid arthritis
    c. Septic arthritis
    d. Gout

28. A lateral curvature of the spine, often treated in childhood is called:
    a. scoliosis.
    b. lordosis.
    c. kyphosis.
    d. osteoporosis.

29. Before a bone can heal, it must be _____ to make sure the ends of the bone are touching.
    a. immobilized
    b. reduced
    c. repaired
    d. rebroken

30. Pins, screws, or plates are used to fix bones in place during _____ reduction.
    a. closed
    b. delayed
    c. surgical
    d. medical

31. The first step in bone repair is:
    a. hematoma formation.
    b. soft callus formation.
    c. bony callus formation.
    d. remodeling.

32. When should you see a doctor for joint pain?
    a. When signs of infection appear
    b. When there is obvious deformity
    c. When there is no improvement after 7 days
    d. All of the above

33. Which of the following is a congenital bone disorder?
    a. Osteogenesis imperfecta
    b. Osteoporosis
    c. Fracture
    d. Osteomyelitis

34. This type of fracture, in which the bone is splintered, is common in patients with osteoporosis:
    a. compound.
    b. spiral.
    c. simple.
    d. None of the above

35. This disorder is a common cause of lumbar pain:
    a. scoliosis.
    b. herniated disk.
    c. fracture.
    d. traction.

MATCHING EXERCISES

Set 1

Please match each common name with the medical term for that bone(s).

_____ 1. True ribs
_____ 2. Shoulder blade
_____ 3. Upper arm bone
_____ 4. Fingers and toes
_____ 5. Thigh bone
_____ 6. Lower leg bone
_____ 7. Forearm bone
_____ 8. Wrist bones
_____ 9. Ankle bones
_____ 10. Foot bones

a. Clavicle
b. Metatarsal
c. Scapula
d. Radius
e. Femur
f. Vertebrocostal
g. Fibula
h. Tarsals
i. Phalanges
j. Humerus
k. Carpals
l. Metacarpals
m. Vertebrosternal

Set 2

Please match each term with the appropriate location and movement.

_____ 1. Sutures
_____ 2. Synovial
_____ 3. Ball and socket
_____ 4. Gliding
_____ 5. Saddle
_____ 6. Pivot
_____ 7. Condyloid
_____ 8. Cartilaginous
_____ 9. Fibrous
_____ 10. Hinge

a. Neck and forearm; rotates
b. Found at the pubic symphysis and joining ribs to sternum, little movement
c. Hips and shoulder; multiple movement
d. Knees and elbow; allows for flexion and extension
e. Found on the cranium, no movement
f. Base of the thumb; multiple movement including opposition
g. Fluid in a joint cavity, lubrication
h. Holds ulna and radius together; little movement
i. Found between the carpals and between the tarsals
j. Found between wrist bones and the forearm bones; allows biaxial movement

Set 3

Please match each term with the appropriate definition.

_____ 1. Diaphysis
_____ 2. Facet
_____ 3. Tubercle
_____ 4. Fossa
_____ 5. Meatus
_____ 6. Sinus
_____ 7. Crest
_____ 8. Head
_____ 9. Foramen
_____ 10. Spine

a. A tube or tunnel-like passageway through a bone
b. A nontubular passageway through a bone for ligament, nerves, and blood vessels
c. A small flattened area
d. The shaft of the bone
e. A knoblike projection
f. A sharp pointed projection
g. A hollow area; space within a bone
h. An articulating end of a bone that is rounded
i. A narrow ridge
j. Shallow depression

Set 4

Please match each disorder with the appropriate treatment.

_____ 1. Osteoarthritis
_____ 2. Rheumatoid arthritis
_____ 3. Ligament tears
_____ 4. Gout
_____ 5. Kyphosis
_____ 6. Osteomyelitis
_____ 7. Osteoporosis
_____ 8. Tendonitis
_____ 9. Rickets
_____ 10. Herniated disk

a. Antibiotics, surgery, prevention
b. Anti-inflammatories, steroids, methotrexate, immune suppressants, exercise, surgery
c. Heat, exercise, anti-inflammatories, surgery
d. Calcium supplements, exercise, estrogen replacement
e. Rest, immobilization, surgery
f. Pain medication, rest, heat, ice
g. Vitamin D supplement
h. Rest, heat, exercise, medication, training in lifting techniques, surgery
i. Exercise, physical therapy, bracing
j. Diet, medication, rest, exercise

FILL IN THE BLANK

Fill in the blanks to complete the following statements.

1. The medical condition called _____ is a degenerative disorder characterized by a decrease in bone density.

2. Cleft palate and clubfoot are examples of _____ disorders.

3. Secondary curvatures of the spine are found in the
   _____ and
   _____ vertebral regions.

4. According to your text, osteoclasts arise from
   _____.

5. If you move the joint of the ankle and foot so that you are "standing" on the balls of the foot, you have
   _____ the foot.

6. When a joint is straightened or merely moved so that the angle between the individual bones has increased, the movement is termed _____
   _____.

7. The bone called the _____ is commonly known as the lower jaw.

8. During CPR chest compressions, the part called the
   _____ of the sternum takes the brunt of the compressive force.

9. Ribs 8, 9, and 10, commonly called false ribs, can be clinically called
   _____.

10. Similar to the elbow joint, interphalangeal joints are
    _____ joints.

11. In the creation of the skeletal bones, when shaped cartilage is replaced by osseous tissue, this process is known as
    _____.

12. Where bursitis is inflammation of a bursa, inflammation of the joint is called _____.

13. The human skeleton has _____ bones.

14. Falling off his skateboard, Hugh suffered a(n)
    _____ fracture due to the bones of his forearm being crushed to the point of splintering.

15. Specialized cells called _____ are needed to tear down bone.

16. As people age, bone mass _____.

17. Serious decrease in bone density with age is called
    _____. It greatly increases the risk of fractures.

18. _____ is inflammation of joints, due to autoimmune attack.

19. A(n) _____ is stretching or tearing of a ligament.

20. _____ is minimally invasive joint surgery using a long tube, a light source, and an eyepiece.

21. A herniated _____ often causes back pain by pushing on a spinal nerve root.

22. A fracture in which the bone pushes through the skin is a(n)
_____.

23. An incomplete break, a(n) _____
fracture, is common in children.

24. The last step in bone repair is
_____.

25. _____ is a bone infection.

SHORT ANSWER

1. What are four functions of the skeleton?
_____
_____
_____
_____

2. Compare and contrast the four types of bones.
_____
_____
_____
_____

3. What are the functions of the periosteum?
_____
_____
_____
_____

4. Discuss the difference between a ligament and a tendon.
_____
_____
_____
_____

5. Why are ribs 11 and 12 called floating ribs?
_____
_____
_____
_____

6. What does RICE mean?

_____

_____

_____

_____

7. Compare and contrast rheumatoid arthritis and osteoarthritis.

_____

_____

_____

_____

_____

LABELING ACTIVITIES

1. Label and color the parts of the bone indicated below using Figure 6-3 from your textbook as a guide.

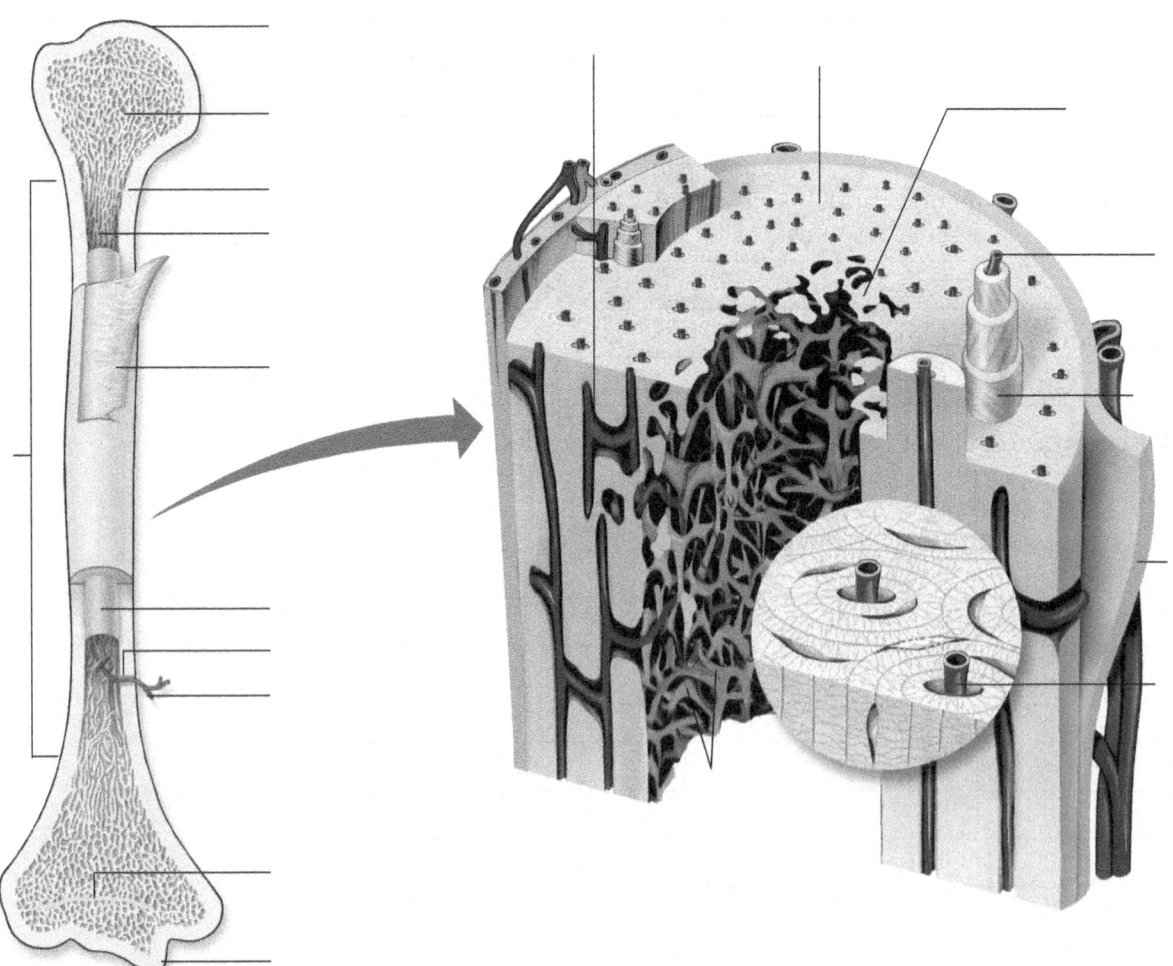

2. Label the parts of the synovial joint using Figure 6-6 from your textbook as a guide.

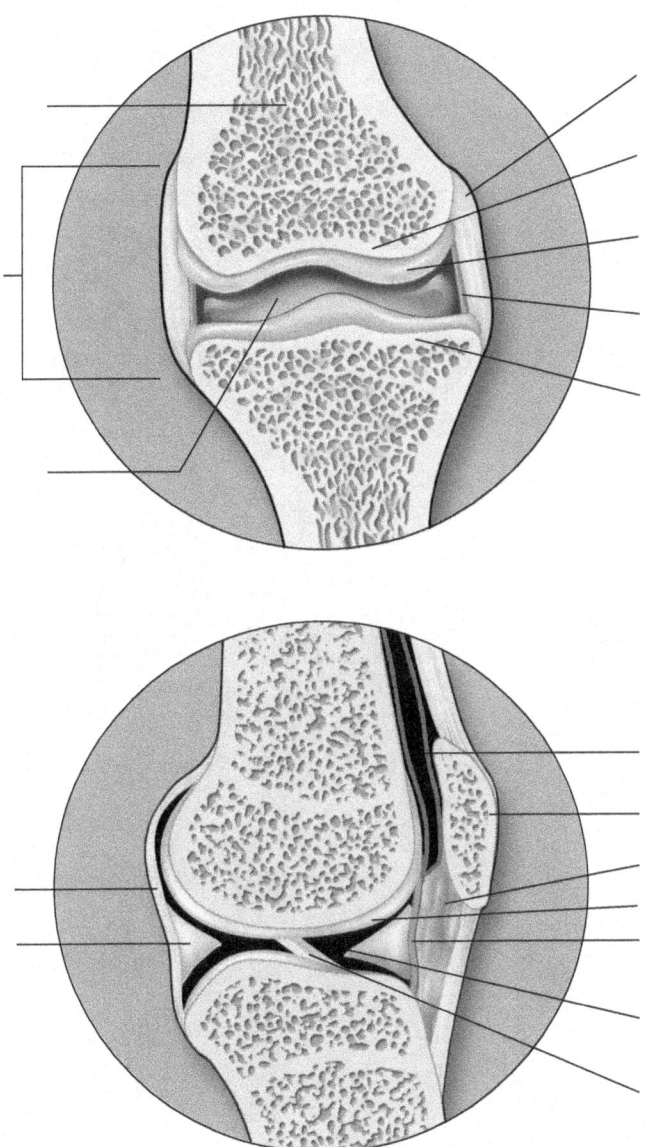

3. Label the bones of the skeleton using Figure 6–9 from your textbook as a
    guide.

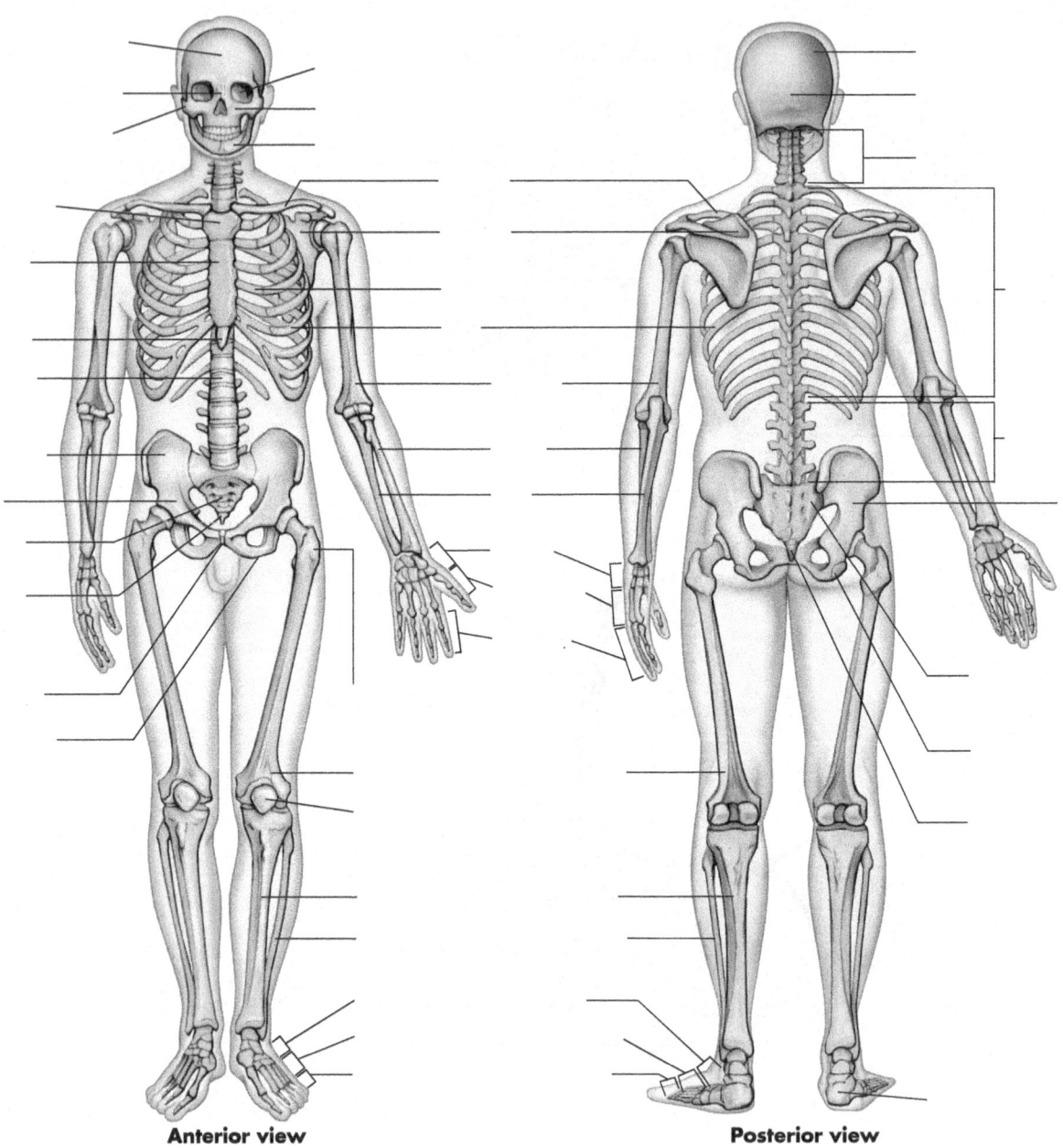

Anterior view   **Posterior view**

CASE STUDY

Gwendolyn owns a home catering business that specializes in good down-home cooking, her favorite kind of food. She has never met a gravy or sauce she couldn't duplicate. Country fried steak with sausage gravy is her specialty. In addition to entrees, she is the caterer of choice for desserts, the richer the better. Lately she has been having pain in her feet, particularly her big toes, which are red and swollen. After a couple of weeks of icing and over-the-counter medication with no improvement, she makes an appointment with her physician.

1. Without ordering even one test, her physician has a preliminary diagnosis. What is it?

   _____

   _____

   After X-rays and blood tests, the diagnosis is confirmed. She has not injured her toes, nor does she have rheumatoid or osteoarthritis. Her doctor's first guess was right. When the doctor tells Gwendolyn the treatment she is aghast. She can't do that!

2. What is the primary treatment for her disorder?

   _____

   _____

3. What might her doctor prescribe to decrease the chance of her symptoms returning?

   _____

   _____

LEARNING ACTIVITIES

1. Get a cutout of a skeleton, like a Halloween decoration. How many bones can you name?

2. Take a survey of the students in the class or your friends and the members of your family. How many have had bone and/or joint injuries? What was the treatment? Make a table of types of injuries and treatments.

3. Buy a cheap plastic skeleton from a hobby shop or card store. Take it apart. Can you put it back together?

4. With a partner, demonstrate joint movements. Can you tell which is which? See who can make the correct movement when "ordered" to do it by another student.

5. For each joint, figure out which movements it can make. Does the list make sense given the structure of the joint?

THE MUSCULAR SYSTEM:
MOVEMENT FOR THE JOURNEY

CHAPTER SUMMARY

Muscle tissue is tissue that can contract, or shorten, with power. It is the tissue that gives us our ability to move. There are three types of muscle tissue: skeletal, attached to bones; visceral, in walls of hollow organs and tubes; and cardiac, in the heart wall. Skeletal muscle is under conscious control and is therefore known as voluntary muscle. It controls the movement of your skeleton. Both smooth and cardiac muscle are not under conscious control and are considered involuntary. Cardiac and smooth muscle move organs. Cardiac muscle causes the heart to pump blood, and smooth muscle causes the movement of other organs, tubes, and vessels. Cardiac and skeletal muscle are called striated muscle because their structure causes them to appear striped. Smooth muscle gets its name because it is not striated. Normally muscles exhibit muscle tone, which is partial contraction. With increased use, muscles may increase in size (hypertrophy). With decreased use, muscles may waste away (atrophy). Myopathy, muscle disease or disorder from many different causes, may cause atrophy.

Skeletal, or voluntary, muscle is attached to bones by tendons or aponeuroses. Damage to this connective tissue or to the muscle itself is a muscle strain. The most severe strains actually result in the rupture (complete tearing) of muscles or tendons. Chronic damage to tendons is called tendonitis or tendinosis. In your body there are dozens of muscles, each with unique actions, or movements. They are named based on a series of rules that tell you their action, location, attachment site, or shape. There are unique names for every muscle and a series of terms to describe specific muscle movements. Fibromyalgia syndrome is the name for a disorder characterized by extreme muscle tenderness and fatigue with other symptoms.

The movement of a skeletal muscle can be further characterized based on its relation to other muscles and/or based on the change the muscle causes in the position of the joint moved when the muscle contracts. No matter the type of movement, all skeletal muscles cause movement by contracting. Contraction is caused by changes in the position of tiny protein rods in muscle cells. These rods (myofilaments) are organized into units called sarcomeres. It is the sarcomeres that actually get shorter when muscles contract.

Skeletal muscles are composed of cylinders within cylinders within cylinders. A muscle is made up of bundles of muscle fibers (muscle cells). Each muscle fiber contains several cylindrical myofibrils. Inside the myofibrils are the myofilaments. Muscle fibers also contain sarcoplasmic reticula (modified endoplasmic reticula), sarcolemma (specialized cell membranes), and dystrophin (special protein). Defective dystrophin leads to a fatal genetic disease known as Duchenne muscular dystrophy.

Muscles contract when the chemical acetylcholine is released from a motor neuron at the neuromuscular junction. Acetylcholine binds to the muscle, opening sodium channels, and exciting the muscle. As the muscle becomes excited, calcium is released from the sarcoplasmic reticulum. The calcium allows the myofilaments to bind to each other and ultimately shorten the sarcomeres. ATP is necessary for the binding. Myasthenia gravis, which results in progressive muscle weakness, is caused by the destruction of acetylcholine receptors on muscle cells. To move, muscles need fuel in the form of ATP. To prolong contraction, many muscles have stored glycogen and/or fat, or the ability to contract without oxygen for short periods of time.

Smooth muscle and cardiac muscle are involuntary muscles. Smooth muscle, found in the walls of tubes and hollow organs, is not striated. Cardiac muscle, found in the heart wall, is striated.

CHAPTER OUTLINE

MEDICAL TERMINOLOGY REVIEW

Define the following terms.

1. Myopathy: _____

2. Myalgia: _____

3. Paralysis: _____

4. Hypertrophy: _____

5. Atrophy: _____

6. Tonus: _____

7. Spasm: _____

8. Autoimmune disorder: _____

9. Electromyogram: _____

10. Tonus: _____

MULTIPLE CHOICE

Circle the letter of the correct answer.

1. Choose the correct structural arrangement from macro to micro in terms of size.
   a. Muscle cells, myofibrils, sarcomere, myofilament
   b. Myofibril, myofilament, muscle cell, sarcomere
   c. Myofilament, muscle cells, myofibrils, sarcomere
   d. Muscle cell, myofilament, sarcomere, myofibrils

2. Which of the following is a group of anterior thigh muscles?
   a. Hamstrings
   b. Quadriceps
   c. Peroneals
   d. Gluteals

3. Which of the following is a group of buttocks muscles?
   a. Psoas
   b. Gluteals
   c. Hamstrings
   d. Quadriceps

4. Which of the following is a muscle of the lower leg?
   a. Gastrocnemius
   b. Latissimus dorsi
   c. Deltoid
   d. Hamstring

5. Muscles that are used for duration or high endurance activity will:
   a. look white due to the excess oxygen and fat stored for energy.
   b. look white due to lack of blood supply.
   c. look dark due to the rich blood supply to carry needed oxygen.
   d. look dark due to chronic tears and scar tissue in the muscle fibers.

6. Where are calcium ions stored in the muscle cells?
   a. End bulb
   b. Nucleus
   c. Myosin cross bridges
   d. Sarcoplasmic reticulum

7. Which of the following muscles is under voluntary control?
   a. Skeletal
   b. Cardiac
   c. Visceral
   d. Smooth

8. After death, when the body becomes stiff due to unreleased muscle contraction, the condition is referred to as:
   a. rigor mortis.
   b. tetanus.
   c. myalgia.
   d. paralysis.

9. When the diameter of a blood vessel increases:
   a. the pressure also increases.
   b. it is termed vasoconstriction.
   c. the pressure decreases.
   d. Both b and c

10. An injury to a ligament:
    a. strain.
    b. sprain.
    c. staine.
    d. pull.

11. Which of the following correctly describes an aponeurosis?
    a. A psychosis in which pain is felt in an area of a limb that has been amputated
    b. Inability to move the neck muscles
    c. A flat, broad, tendonlike sheath
    d. Necrosis (death) of the muscle cells

12. Why are migratory birds' breasts dark (as in the king eider, arctic tern, and blue winged teal) and nonmigratory birds' breasts white (as in the turkey, sandhill crane, and red cardinal)?
    a. Migratory birds need speed to traverse far distances; dark meat is a clear attribute for speed.
    b. Nonmigratory birds need endurance to traverse far distances; dark meat is a clear attribute for endurance.
    c. Migratory birds need endurance to traverse far distances; dark meat is a clear attribute for endurance.
    d. Nonmigratory birds need speed to traverse far distances; white meat is a clear attribute for endurance.

13. What is/are energy source(s) used by muscle?
    a. Calcium
    b. Fat
    c. Glucose
    d. Both b and c

14. Which of the following is true about the sliding filament theory and consequently about muscle contraction?
    a. Cross bridges are formed between actin and myosin; myosin rotates, pulling the actin toward the center of the sarcomere.
    b. Cross bridges are formed between actin and the Z-lines; Z-lines rotate, and as a result, myosin shortens.
    c. Cross bridges are formed between actin and myosin; actin rotates, pulling myosin toward the Z-lines, shortening the sarcoplasmic reticulum.
    d. Sarcoplasmic reticulum releases phosphorous, resulting in cross bridges forming between myosin and the Z-lines; actin rotates, pulling toward the center of the sacromere.

15. Some sphincters are examples of:
    a. cardiac muscles.
    b. smooth muscles.
    c. visceral muscle.
    d. Both b and c

16. If the erector spinae muscles are the antagonist, which of the following will be a prime mover?
    a. Latissimus dorsi
    b. Trapezius
    c. Rectus abdominis
    d. Both a and b

17. One of the calf muscles, called the soleus, when contracted moves the heel of the foot (calcaneus) closer to the posterior leg. Given this information and your knowledge of the principles of origin and insertion, what is the muscle's origin?
    a. Calcaneus
    b. Anterior leg
    c. Posterior leg
    d. Both a and b

18. A group of muscles called the scalenes laterally flexes the neck. Given this information and your knowledge of the principles of origin and insertion, which of the following most likely is its insertion?
    a. Cervical vertebrae
    b. Shoulder
    c. Ribs
    d. Collarbone

19. Besides a physical separation of the thoracic cavity and the abdominal cavity, what purpose does the diaphragm serve?
    a. Flexes the trunk
    b. Extends the trunk
    c. Controls breathing
    d. All of the above

20. The diaphragm is what type of muscle?
    a. Smooth
    b. Cardiac
    c. Visceral
    d. Skeletal

21. The diaphragm is under what type of control?
    a. Voluntary
    b. Involuntary
    c. Both voluntary and involuntary
    d. Neither voluntary or involuntary

22. A muscle called the deltoid pulls the arm away from the body, directly out away from the sides. This movement is referred to as:
    a. rotation.
    b. abduction.
    c. adduction.
    d. lateral flexion.

23. Which of the following muscles are striated?
    a. Sphincters
    b. Walls of blood vessels
    c. Muscles that move the upper arm
    d. Muscles of peristalsis

24. When the diameter of a blood vessel decreases:
    a. the pressure also decreases.
    b. it is termed vasoconstriction.
    c. the pressure increases.
    d. Both b and c

25. Muscles that extend the forearm at the elbow most likely will have their bellies (bulging part) located in the:
    a. anterior forearm.
    b. anterior arm.
    c. posterior forearm.
    d. posterior arm.

26. A _____ strain will result in severe pain, extensive bruising, swelling. and loss of movement.
    a. mild
    b. moderate
    c. severe
    d. ligament

27. _____ is a degenerative disease leading to breakdown and scarring of tendons.
    a. Strain
    b. Tendinitis
    c. Tendinosis
    d. Both b and c

28. Which of the following may be a cause of fibromyalgia syndrome?
    a. Hyperactive stress response
    b. Hypochondria
    c. Mental illness
    d. Exercise

29. The breakdown of muscle fibers is the chief problem in Duchenne muscular dystrophy due to a genetic mistake in the gene for this protein.
    a. Actin
    b. Myosin
    c. Dystrophin
    d. Contractin

30. A bacterial toxin due to a wound infection causes tetanus and has the following symptom.
    a. Muscle weakness
    b. Numbness
    c. Severe muscle spasm
    d. Decreased muscle tone

31. The P in "PRICE" stands for:
    a. pain relief.
    b. protection.
    c. pleasant.
    d. prevention.

32. Which of the following is a potential cause of myopathy?
    a. Genetics
    b. Nervous system problems
    c. Injury
    d. All of the above

33. Which of the following is a type of strain?
    a. Pulled muscle
    b. Torn ligament
    c. Torn ACL
    d. Herniated disk

34. Why are diabetic patients at increased risk of tendon injuries?
    a. They are out of shape
    b. They do not exercise enough
    c. They have impaired wound healing
    d. Genetics

35. What muscle change is a serious risk for patients who are confined to bed?
    a. Atrophy
    b. Muscular dystrophy
    c. Tetanus
    d. Tendonitis

 ## MATCHING EXERCISES

Set 1

Please match each muscle with its location.

_____ 1. Latissimus dorsi
_____ 2. Pectoralis major
_____ 3. Rectus abdominis
_____ 4. Erector spinae
_____ 5. Orbicularis oculi
_____ 6. Masseter
_____ 7. Orbicularis oris
_____ 8. Mentalis
_____ 9. Sternocleidomastoid
_____ 10. External obliques

a. Muscle encircling the mouth
b. Muscle encircling the eyes
c. Muscle to the side of the jaw
d. Neck muscle
e. Chest muscle
f. Vertical muscle from inferior margin of rib cage to the pubis
g. Lateral abdominal muscle
h. Back muscle running from the vertebrae to the upper arm
i. Vertical back muscle running from lower vertebrae to upper vertebrae
j. Muscle of the mid chin

Set 2

Please match each term with the appropriate definition.

_____ 1. Flexion
_____ 2. Rotation
_____ 3. Abduction
_____ 4. Extension
_____ 5. Adduction
_____ 6. Vasodilation
_____ 7. Vasoconstriction
_____ 8. Tetany
_____ 9. Antagonist
_____ 10. Agonist

a. Prime mover
b. Lengthens on movement or contraction of prime mover
c. Movement away from midline
d. Movement toward midline
e. Movement decreasing angle of the joint
f. Movement increasing the angle of the joint
g. Movement decreasing the diameter of the blood vessel
h. Movement increasing the diameter of the blood vessels
i. Movement around a center axis
j. Movement that creates rigid paralysis

Set 3

Please match each disorder with the appropriate desccription.

_____ 1. Myalgia
_____ 2. Hernia
_____ 3. Strain
_____ 4. Cramp
_____ 5. Sprain
_____ 6. Myasthenia gravis
_____ 7. Guillain-Barré syndrome
_____ 8. Botulism
_____ 9. Atrophy
_____ 10. Muscular dystrophy

a. Tear or injury in muscle and/or tendon
b. Involuntary, sudden, and violent contractions
c. Tears or breaks in a ligament
d. A PNS disorder resulting in flaccid paralysis
e. An autoimmune disorder that disrupts synaptic transmission
f. Inherited muscle disease in which muscle fibers degenerate
g. A potentially deadly disease that causes paralysis and is a result of ingested bacteria
h. Tenderness and pain in muscle
i. A tear in a muscle wall through which an organ protrudes
j. Condition marked by rigid muscle spasm caused by a bacteria most likely entering the body via a puncture wound
k. The process of muscle wasting away; could be due to lack of nutrition, disease, or disuse

Set 4

Please match each disorder with the appropriate treatment.

_____ 1. Torn tendon
_____ 2. Tendinosis
_____ 3. Fibromyalgia syndrome
_____ 4. Duchenne muscular dystrophy
_____ 5. Myasthenia gravis
_____ 6. Tetanus
_____ 7. Shin splints
_____ 8. Spasm
_____ 9. Mitochondrial myopathy
_____ 10. Pulled muscle

a. AchE inhibitors, steroids, immunosuppressants
b. No treatment to stop progression
c. Rehydration, rest, stretching
d. No real treatment, manage symptoms
e. PRICE, PT, medication
f. Antibiotics, ventilation, pain management
g. Antidepressants, antiepileptics, pain relief
h. Medication, ice, rest, new footwear
i. Surgery followed by PT
j. Ice, rest, gradual resumption of activity

FILL IN THE BLANK

Fill in the blanks to complete the following statements.

1. Structures called _____
    allow for uniform contraction of the cardiac muscle.

2. The rhythmic internal movement of food though the GI tract is termed
    _____.

3. The _____ muscle is the
    antagonist of the biceps brachii.

4. A muscle called the palmaris longus pulls the hand closer to the forearm. This action at the wrist is known as

_____.

5. To generate heat, the body _____,
a biological reaction figuratively saying it is too cold.

6. To supply energy and heat, the body converts stored
_____ into glucose.

7. The functional unit of the muscle is the

_____.

8. Rigor mortis occurs when _____
stays in the cytoplasm.

9. Neurons secrete a neurotransmitter called
_____, which sets the
process of muscle contraction in motion.

10. Each functional unit of the muscle is separated from others by

_____.

11. Provided that the hamstrings are the prime movers, the
_____ are the antagonists.

12. The group of muscles called the
_____ assists the prime
movers in a particular movement.

13. Smooth muscle is also called
_____ muscle.

14. Cardiac muscle forms the walls of the

_____.

15. A fibrous tissue attaching bone to bone is a(n)

_____.

16. _____ is the general term
for muscle disease or disorder.

17. A strain is an injury to a(n)

_____.

18. _____ is a chronic pain
syndrome characterized by bilateral tenderness, fatigue, sleep disorders,
depression, and exercise intolerance.

19. Muscle pain or tenderness is also known by this medical term

_____.

20. _____ is a genetic
myopathy found overwhelmingly in boys.

21. Myasthenia gravis is a(n) _____
disorder causing progressive muscle weakness.

22. Though easily prevented by vaccination, _____, caused by a bacterial infection, is a common serious disease in developing countries.

23. Malfunction of the _____, a cellular organelle, causes several forms of incurable genetic myopathy.

24. Pain in the tibial region is a symptom of _____, a common tendon injury in runners.

25. Treatment choice for myopathy depends on the _____.

SHORT ANSWER

1. Why does smooth muscle form scars rather than effectively healing itself?

_____

_____

_____

_____

2. Besides movement, what are two other very important functions of the muscular system?

_____

_____

_____

_____

3. Describe four places where smooth muscle can be found in the body.

_____

_____

_____

_____

4. What gives skeletal muscle its striped appearance?

_____

_____

_____

_____

5. List and describe the three types of muscle tissue.

   _____

   _____

   _____

   _____

6. Why is tendonitis not the best terminology to describe chronic tendon
   injuries?

   _____

   _____

   _____

   _____

7. List the possible causes of myopathy and one example for each cause.

   _____

   _____

   _____

   _____

LABELING ACTIVITIES

1. Label the muscles using Figure 7–2 in your textbook as a guide.

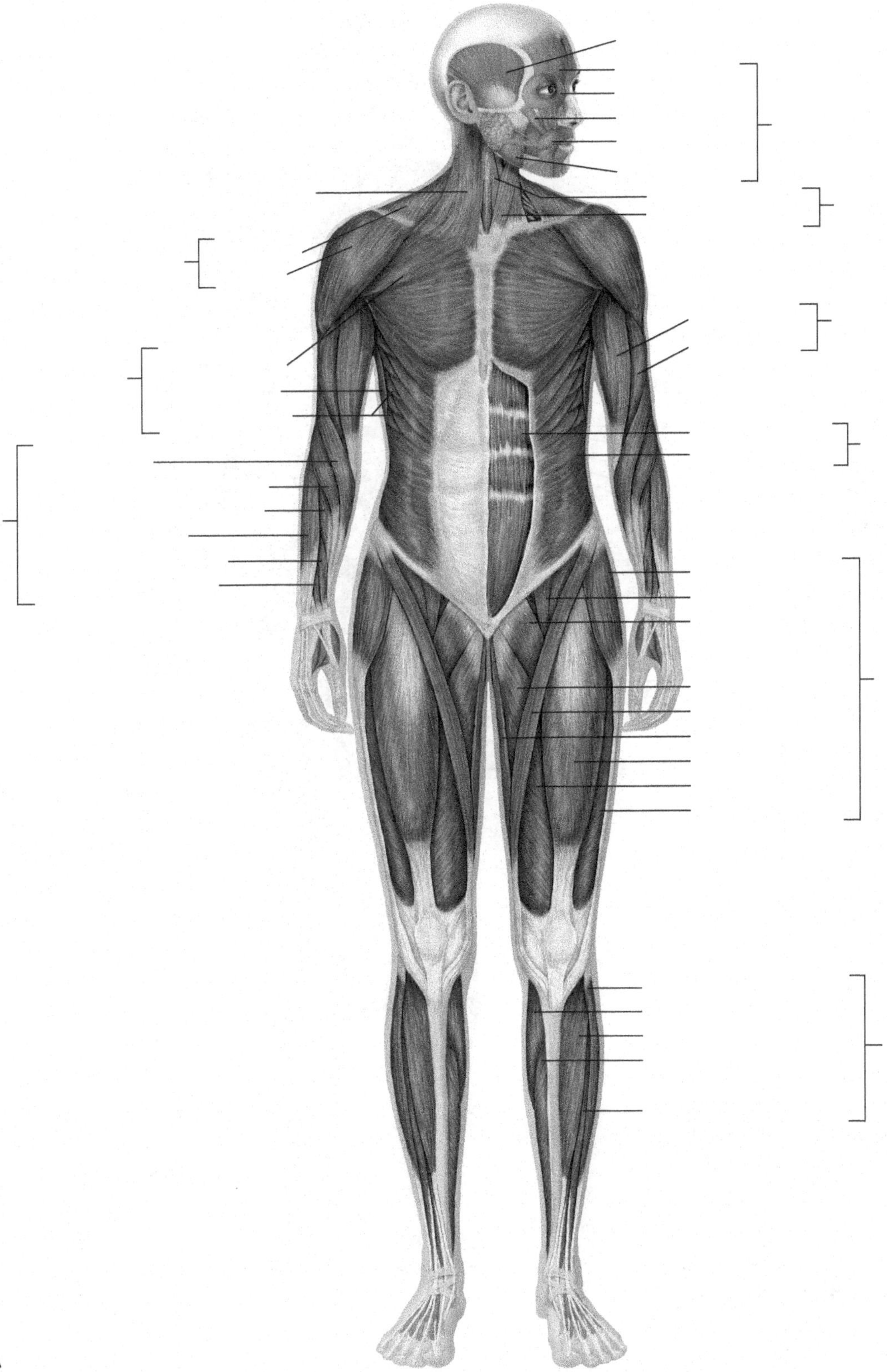

A

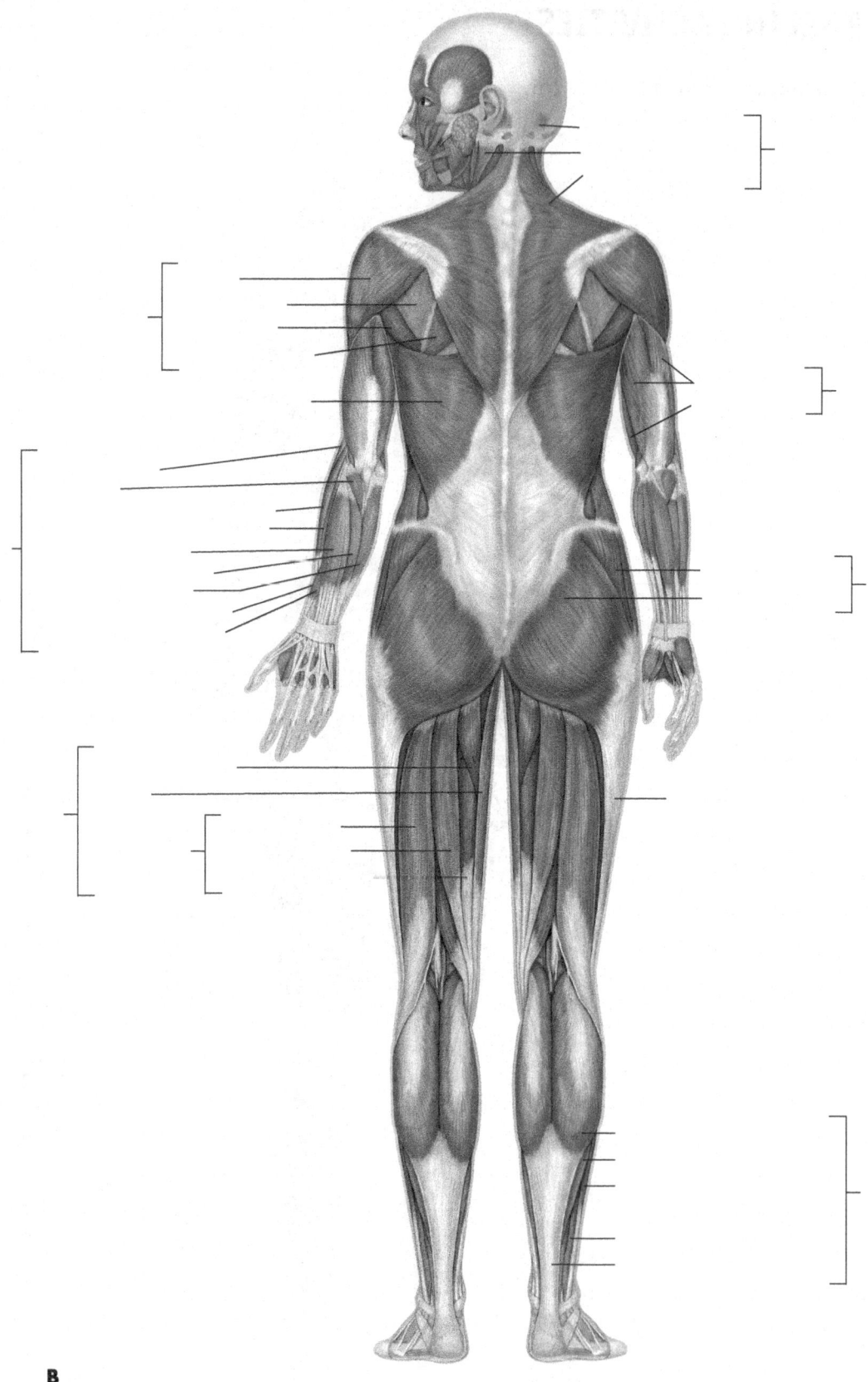

B

2. Label the parts of the muscle and muscle fiber using Figure 7–10 in your textbook as a guide.

A MUSCLE SEGMENT

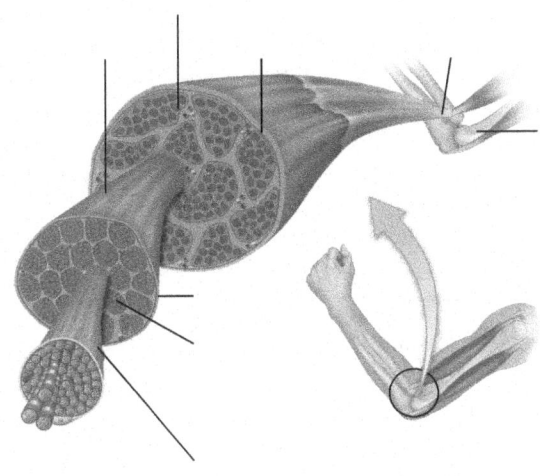

B MUSCLE SEGMENT WITH SARCOMERE

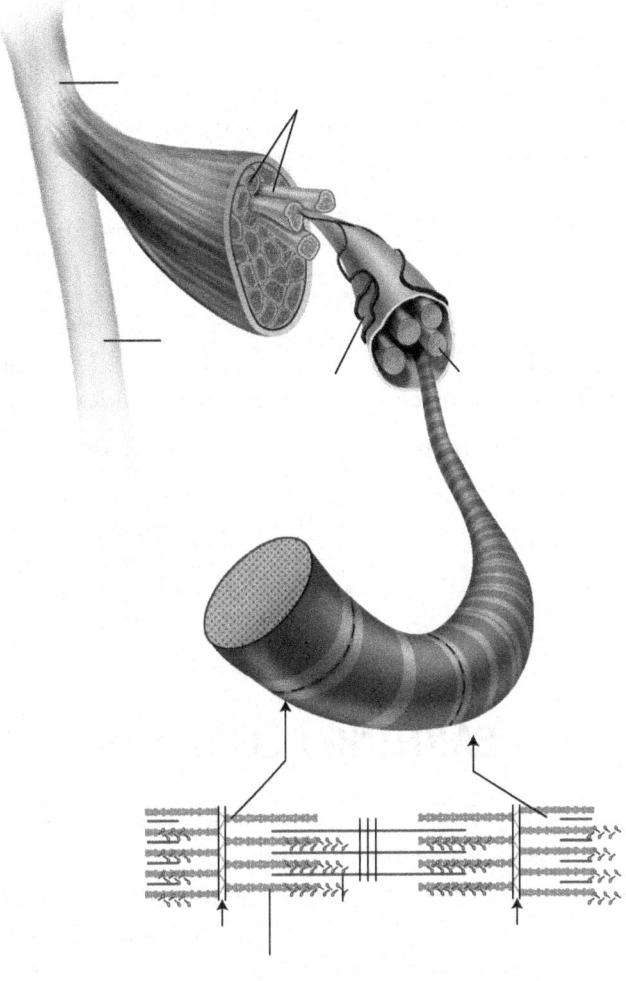

C SARCOMERE

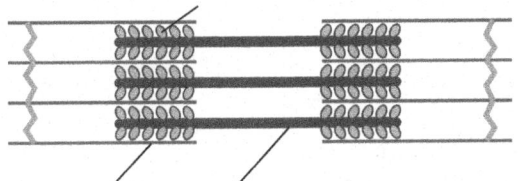

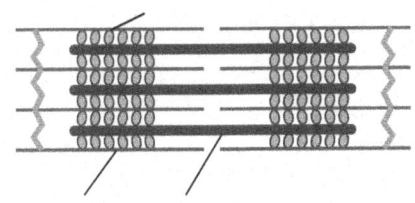

CASE STUDY

Bill, playing his usual Sunday afternoon basketball game, ran downcourt, leapt into the air, and snagged the ball, stuffing it into the hoop. Instead of a graceful landing, he collapsed to the ground, writhing and holding his right ankle. His buddies carried him off the court and whisked him away to the nearest emergency department.

Physical examination and x-rays indicated a bad strain or sprain.

1. What treatment was prescribed?

_____

_____

_____

Even after a month of treatment, Bill's ankle is not better. An MRI reveals that Bill has a ruptured (completely torn) Achilles (calcanial) tendon.

2. What treatment is indicated now that the injury has been accurately diagnosed?

_____

_____

_____

LEARNING ACTIVITIES

1. For 10 muscles of your choice, explain how they are named based on the rules for naming muscles.

2. On a plastic or cardboard skeleton, use yarn to simulate muscle action. For the major muscles, attach one end to the insertion and pull it toward the origin. How do the origin and insertion relate to the action?

3. There are many different myopathies with many different causes. Using the Internet, compile a list of myopathies and their underlying causes. What kinds of problems can lead to muscle disease?

4. Play muscle "Jeopardy." One person lists an action, attachments, location etc., and the other identifies the muscle described. The answer should be in the form of a question, of course!

5. There are many muscle-enhancing dietary supplements on the market. Using the Internet, investigate one of these supplements. Is there any biological basis for the product's claims?

THE INTEGUMENTARY SYSTEM: THE PROTECTIVE COVERING

CHAPTER SUMMARY

The integumentary system's only organ is your skin, the largest organ in your body and the major barrier between your body and the outside world. The skin has several functions including protection against infection, prevention of fluid loss, protection against the sun, fat storage, production of vitamin D, touch sensation, and regulation of body temperature. The skin is divided into three major layers, the epidermis, the dermis, and the hypodermis (subcutaneous fascia). The epidermis consists of keratinized stratified squamous epithelium. The cells in the epidermis are born in the deepest layer of the epidermis and move toward the surface. As they move toward the surface, they die and become filled with a hard protein called keratin. The very top layer of the skin, then, is dead. The epidermis is divided into several layers. Skin color is determined by the pigmentation of the epidermis, though some diseases can cause changes in skin color.

The dermis is a connective tissue layer that supports the epidermis. It consists of connective tissue with collagen and elastic fibers, as well as nerves, blood vessels, and accessory structures such as hair follicles and oil and sweat glands. The subcutaneous fascia, or hypodermis, is a fatty layer under the dermis that protects, insulates, and allows attachment of skin to the rest of the body. Though skin acts as a pretty effective barrier for infection, the skin itself can become infected by many different pathogens, including viruses, bacteria, protozoans, and fungi. Many skin infections are chronic and/or difficult to treat.

When skin is damaged, whether from injury, disease, or infection, it has an extraordinary ability to heal. The wound is first plugged by platelets. Then inflammation sets in, and the wound is patched by collagen fibers. Finally, tissues may be replaced by the original tissue or the patch becomes a permanent scar. The healing ability of skin may be seriously challenged when skin is burned. Burns, whether from chemicals, heat, or radiation, destroy tissue. The severity of a burn is determined by its depth (degree) and extent (percent of body). The deeper and/or more widespread the burn, the more serious. Very deep or extensive burns are life-threatening because patients lose fluid, have difficulty regulating temperature, and are susceptible to infection.

The skin has several accessory organs, including nails and hair. Nails and hair are both made of keratin, the same protein that is in the epidermis. Thus, nails and hair are both dead, though they are made by living cells, nails by the nail root and hair by the follicle. A number of disorders cause changes in the appearance of nails or hair. Lice and scabies are insects that infest skin. Lice are found associated with body hair.

Skin is important in temperature regulation. Body temperature is controlled by the amount of blood flowing to the skin. When you are cold, less blood flows to the skin, and when you are warm, more blood flows to the skin.

Many skin disorders are caused by infections, injury, systemic disorders, and allergies. Some are easy to treat and temporary, whereas others are difficult to treat or chronic. See the chapter for a complete description of selected disorders.

CHAPTER OUTLINE

I. Overview
   A. Functions
   B. Skin anatomy
      1. Epidermis
      2. Dermis
      3. Subcutaneous fascia
      4. Pathology connections
         a. Discolorations
         b. Infections

II. How skin heals
   A. Wound healing
   B. Pathology: burns

III. Nails

IV. Hair
   A. Anatomy
   B. Pathology: lice and scabies
   C. Forensics

V. Temperature regulation

VI. Pathology
   A. Types of skin lesions
   B. Table of skin diseases

VII. Pharmacology corner

MEDICAL TERMINOLOGY REVIEW

Define the following terms.

1. Macule: _____

2. Pustule: _____

3. Wheal: _____

4. Ulcer: _____

5. Papule: _____

6. Crust: _____

7. Nodule: _____

8. Scale: _____

9. Vesicle: _____

10. Fissure: _____

 ## MULTIPLE CHOICE

Circle the letter of the correct answer.

1. Hair is composed of a protein called:
   a. hemoglobin.
   b. lanugo.
   c. lunago.
   d. keratin.

2. This skin condition is caused by the herpes simplex virus forming a visible lesion on the lip margins:
   a. psoriasis.
   b. pustule.
   c. cold sore.
   d. acne.

3. Usually observed in children, this communicable disease is caused by mites:
   a. mumps.
   b. chicken pox.
   c. scabies.
   d. hives.

4. "Jock itch" is caused by which of the following?
   a. Fungi
   b. Viruses
   c. Bacteria
   d. Protozoa

5. Some people of European descent have a pinkish tone to their skin because of:
   a. melanin.
   b. carotene.
   c. lack of melanocytes.
   d. surface vascularizaton.

6. Which of the following correctly describes an abrasion?
   a. Raised skin, pimples with a defined border
   b. Open, craterlike sore where tissue death is evident
   c. Scratched off skin
   d. Bruise

7. Acne develops when:
   a. glucose is consumed and sweets solidify in the sebaceous glands.
   b. there is an overproduction of sebum and an inflammation of the oil glands.
   c. sweat glands are blocked with dirt and environmental grime.
   d. sweat becomes too sweet, and viruses are attracted and accumulate at the pore opening.

8. In a cold environment, blood vessels of the skin:
   a. vasoconstrict, channeling blood from the periphery to the core to collect more blood cells and fibroblasts.
   b. vasodilate, accepting more blood from the body's core to be warmed by solar energy.
   c. vasoconstrict, channeling blood from the periphery toward the body's core where heat is.
   d. vasodilate, accepting blood from the body's core to dissipate heat at the skin's surface.

9. To which of the following structures do the arrector pili muscles attach?
   a. Hair follicle
   b. Stratum corneum
   c. Elastin
   d. Subcutaneous fascia

10. Based on the information given in Chapter 8, for effective body cooling to occur:
    a. water from sweat glands is excreted onto the skin, then is evaporated off, dispelling heat from the surface.
    b. water from sweat glands is excreted onto the skin, then as soon as it is felt, needs to be wiped off, carrying heat with the towel or cloth used.
    c. urination needs to cease, and excessive water consumption must be temporarily stopped.
    d. sweat needs to be at the body's core temperature, and nitrogenous wastes must be absent in the sweat excretion.

11. What determines the texture of hair?
    a. Shaft shape: flat shafts produce curly hair, and round shafts produce straight hair.
    b. Follicle shape: vertical follicles produce straight hair, and angular follicles produce curly hair.
    c. Pigmentation: higher concentrations of pigmentation produce curly hair.
    d. Heat and humidity: people exposed to cold will grow straight hair, and people exposed to heat will not.

12. What determines the color of hair?
    a. Carotene
    b. Melanin
    c. Keratin
    d. Bilirubin

13. Why should you not squeeze blackheads?
    a. It may create a pit.
    b. A substance in the pores and glands is highly contagious.
    c. It may force the infection back into the sudoriferous pore.
    d. All of the above

14. What is/are the danger(s) of washing the face too often with non-pH-balanced soap?
    a. Loss of pigmentation
    b. Decrease blood supply
    c. Loss of nerve sensation
    d. Loss of antibacterial barrier

15. Shaving or frequent trimming will:
    a. cause hair to grow faster.
    b. cause hair to grow back slower.
    c. cause hair to grow back curlier.
    d. None of the above

16. What are the three parts of hair?
    a. Cuticle, corium, and papilla
    b. Villa, erector, and lunago
    c. Shaft, body, and cuticle
    d. Follicle, root, and shaft

17. Why is vitamin D a necessity for healthy bones and teeth?
    a. It is needed for the differentiation of osteoclasts to osteoblasts.
    b. It is needed for calcium absorption in the intestine.
    c. It is needed for fighting gingivitis and calcium buildup.
    d. It is not a necessity.

18. What degree is a sunburn?
    a. First
    b. Second
    c. Third
    d. Fourth

19. Which of the following layers of skin is the deepest?
    a. Hypodermis
    b. Dermis
    c. Corium
    d. Epidermis

20. Which of the following cells pull the edges of a wound together?
    a. Red blood cells
    b. Melanocytes
    c. Fibroblasts
    d. Osteocytes

21. What role do white blood cells have in wound healing?
    a. Clotting and secreting meshlike barrier
    b. Dissolving debris by chemically breaking bonds and mechanically pushing debris to the surface
    c. Fighting infection
    d. Blood thinning

22. Which of the following statements is true about melanin, melanocytes, and skin color?
    a. Adult humans, despite race or gender, have the same amount of melanocytes per skin square inch.
    b. Different skin colors and tones are due to different amounts and arrangements of melanocytes.
    c. The more melanin produced, the lighter the skin.
    d. Melanocyte absolute numbers are inversely proportional to the concentration of melanin in the skin; in other words, the more melanocytes, the less pigment can be secreted and can ultimately survive in the skin.

23. A clinician can estimate the extent of the area covered by a burn using what strategy?
    a. Rule of size
    b. Rule of thumb
    c. Rule of nines
    d. Color rule

24. In a condition called cirrhosis, a liver dysfunction, fair skin appears yellow, but in dark skin the yellow may not be evident. Where can the yellow color be seen?
    a. Eyes
    b. Palms
    c. Teeth
    d. Soles

25. Which of the following sweat glands secrete at the hair follicle of sebaceous glands?
    a. Sebaceous
    b. Apocrine
    c. Eccrine
    d. Creatine

26. _____ is a lifelong skin infection, causing blisters, that has periods of remissions and flare-ups.
    a. Poison ivy
    b. Shingles
    c. Herpes
    d. Athlete's foot

27. A sexually transmitted disease of the skin is:
    a. *Herpes zoster.*
    b. *Herpes simplex 2.*
    c. *Herpes varicella.*
    d. *Tinea cruris.*

28. Papilloma virus causes:
    a. warts.
    b. herpes.
    c. chicken pox.
    d. jock itch.

29. This fungal infection affects mainly men:
    a. *Tinea pedis.*
    b. *Tinea cruris.*
    c. *Tinea unguium.*
    d. *Tinea corporis.*

30. Infection of the subcutaneous tissue by staph bacteria is _____ and may be life threatening.
    a. cellulitis
    b. Lyme disease
    c. shingles
    d. herpes

31. Why is nail color important?
    a. It can change with oxygen level.
    b. It changes color during the menstral cycle.
    c. It can indicate overall health.
    d. All of the above

32. This sexually transmitted disease is caused by insect infestation.
    a. Herpes
    b. Pubic lice
    c. Genital warts
    d. Gonorrhea

33. _____ is a relatively rare but very serious type of skin cancer that may be fatal.
    a. Basal cell carcinoma
    b. Squamous cell carcinoma
    c. Melanoma
    d. Keratoma

34. _____ is a skin condition that may become dangerous for bedridden patients.
    a. Abrasion
    b. Ulcer
    c. Eczema
    d. Hives

35. System allergic reactions often present with this skin condition.
    a. Urticaria
    b. Dermatitis
    c. Keloid
    d. Boil

MATCHING EXERCISES

Set 1

Please match each term with the appropriate description.

_____ 1. Bilirubin
_____ 2. Keratin
_____ 3. Fibroblasts
_____ 4. Lipocytes
_____ 5. Sebum
_____ 6. Carotene
_____ 7. Melanin
_____ 8. Corium
_____ 9. Apocrine
_____ 10. Eccrine

a. Yellow jaundice
b. Cools the body
c. Sexual attractant
d. Fat
e. Oil
f. True skin
g. Found in hair and nails
h. Skin healing
i. Darkening of the skin
j. Natural yellow hue to the skin

Set 2

Please match each structure with the appropriate description.

_____ 1. Cuticle
_____ 2. Dermis
_____ 3. Sebaceous
_____ 4. Lunula
_____ 5. Sudoriferous
_____ 6. Stratum basale
_____ 7. Subcutaneous fascia
_____ 8. Epidermis
_____ 9. Hair follicle
_____ 10. Arrector pili

a. Hypodermis
b. Encloses hair root
c. Layer in which epidermal cells are born
d. Lubricates and moisturizes hair and skin
e. Normally seen layer of skin
f. Contractile tissue associated with hair follicle
g. Covers nail root
h. Contains glands, vessels, collagen, and elastin fibers
i. Excretes water and nitrogenous wastes
j. Proximal, whitish, half-moon part of nail

Set 3

Please match each skin lesion with the appropriate description.

_____	1. Freckles	a.	*Herpes zoster*
_____	2. Acne	b.	Patches of excessive melanin production
_____	3. Urticaria	c.	*Herpes simplex*
_____	4. Psoriasis	d.	Cancer
_____	5. Fever blister	e.	Hives
_____	6. Shingles	f.	Open necrotic sore
_____	7. Decubitus ulcer	g.	Itching, scaling, redness, circular borders
_____	8. Abrasion	h.	Bed sores
_____	9. Ulcer	i.	Rubbing off or scratching off of the skin
_____	10. Malignant melanoma	j.	Infection of the sebaceous gland

Set 4

Please match each skin condition with the appropriate treatment.

_____	1. Bed sores	a.	Vaccine, antibiotics
_____	2. Eczema	b.	Treat symptoms, avoid trigger
_____	3. Hives	c.	Surgery, chemotherapy
_____	4. Malignant melanoma	d.	Chemical or physical removal
_____	5. Pediculosis	e.	Clean and dry area, medication
_____	6. Athlete's foot	f.	Prevention, treatment of ulcer
_____	7. Warts	g.	Antihistamines, avoidance
_____	8. Psoriasis	h.	Special soap and shampoo
_____	9. Lyme disease	i.	Medication, UV light
_____	10. Cellulitis	j.	Antibiotics

FILL IN THE BLANK

Fill in the blanks to complete the following statements.

1. Located in the dermis, _____
   fibers help the skin flex with the movement of the body.

2. The most dangerous and life-threatening of the skin cancers is

   _____.

3. Shingles, caused by the

   _____ virus, are found

   mainly on the torso or trunk of the body.

4. The skin muscles that contract, indirectly forming what is
   commonly known as gooseflesh, are clinically called

   _____.

5. The _____ layer
   of the epidermis is constantly shedding as a part of the skin
   replacement   process.

6. The clinical term for pimple is _____.

7. A human adult, having thousands of sweat glands per square inch, has the potential of excreting up to _____ liters of sweat in 24 hours.

8. Normally it takes _____ seconds to reperfuse the nail bed when assessing perfusion of the extremity by squeezing the nail.

9. When a person suffers an injury that does not break the skin yet damages the underlying small blood vessels, this person has suffered a(n) _____.

10. "Dry skin" refers to the lack of _____.

11. The most severe of the burns, _____ degree burn, is marked by tissue damage from the skin's surface to the bone.

12. The patient has suffered _____ percent body-surface-area damage when both right upper and lower limbs, the neck, and the head are burned.

13. Nails grow from the nail _____.

14. The substance or pigment responsible for the darkening of the skin is _____.

15. In hepatitis, a liver disease, _____ builds up in the blood, giving the skin an unhealthy yellowish color.

16. A(n) _____ is a pathologically altered patch of skin.

17. _____ is a bacterial infection of hair follicles.

18. The reawakening of dormant chicken pox years later is known as _____.

19. Plantar warts are found on the _____.

20. Cervical cancer is associated with the _____ virus.

21. Ringworm is actually caused by this type of pathogen: _____.

22. A bull's-eye rash is often the first indication of _____, caused by the bite of an infected tick.

23. A(n) _____ is a scar gone wild.

24. A third-degree burn penetrates through the

_____.

25. _____ is the term for lice
infestation.

SHORT ANSWER

1. How does "gooseflesh" assist in warming the body?

_____

_____

_____

_____

2. Contrast the three types of skin cancers in terms of severity and depth.

_____

_____

_____

_____

3. How do bed sores develop?

_____

_____

_____

_____

4. Besides vitamin D production, what are three functions of skin?

_____

_____

_____

_____

5. What kind of substances can be detected by forensic analysis of the hair?

_____

_____

_____

_____

6. Many skin conditions are caused by infections. List the conditions and
their infectious agents.

_____

_____

_____

_____

7. Discuss the classification of burn severity.

_____

_____

_____

_____

LABELING ACTIVITY

Label and color code the various structures of the skin using Figure 8–1 in your textbook as a guide.

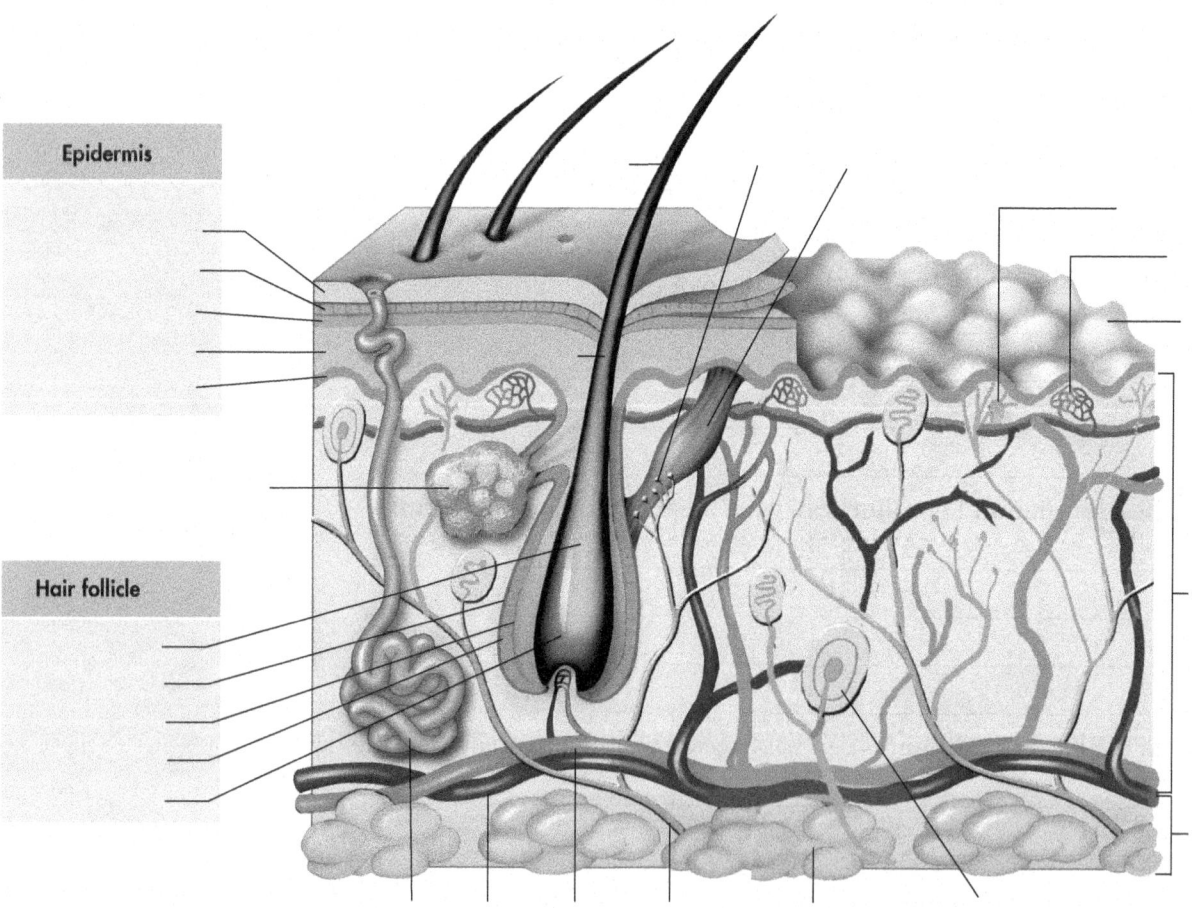

Epidermis

Hair follicle

CASE STUDY

Susie, a 3-year-old, had been playing in the backyard, climbing in and out of the hedges, playing on the swing, and just generally having a good time. It had been a major job to get her to come inside for supper. After supper she began to complain about being itchy. Her mother was shocked to discover that she was covered in some kind of rash. Alarmed, she called the doctor who suggested she take Susie to the emergency department.

By the time they got to the emergency department, Susie was quite ill.

1. After taking a medical history and examining the rash, Susie's pediatrician is pretty sure he knows what is going on. What are the obvious possibilities?

    _____

    _____

2. From the history, the emergency physician discovers that Susie had a peanut butter sandwich for supper and that her best friend has recently had chicken pox. Are there other possibilities for the rash?

    _____

    _____

3. After taking Susie's vital signs and ordering an hour of observation, the doctor gives her an antihistamine, spreads a cream on Susie's rash, and advises her mother to keep Susie out of the hedge if possible. The doctor also suggests oatmeal baths for the itching and assures the worried mother that Susie will be fine though uncomfortable for few days. What is Susie's problem?

    _____

    _____

LEARNING ACTIVITY

1. The textbook could not possibly cover all skin diseases. Use the Internet to research some diseases you haven't read about. What else can go wrong with skin?
2. Many older adults go to great lengths to fight the effects of aging on skin, having face-lifts, dermabrasion, and Botox; coloring their hair to hide gray; or replacing lost hair. Use the Internet to research what happens to skin and hair as people age.
3. Have one student write a scenario in which a patient is burned, describing the depth and extent of the burn. Use the rule of nines and the degree system to diagnose the severity of the burn.
4. There are many accessory structures in the skin. List them and their contribution to skin function. Can you list them all?
5. Play "Name That Disease." One student lists the symptoms and cause of a skin disorder, and other students identify the disorder.

THE NERVOUS SYSTEM: THE BODY'S CONTROL CENTER

Chapter 9

CHAPTER SUMMARY

The nervous system, the body's major control system, can be thought of as a computer, with inputs (sensory), outputs (motor), and a central processing unit (brain and spinal cord). The inputs and outputs are carried out by the peripheral nervous system (PNS), and the brain and spinal cord are the central nervous system (CNS). The sensory parts of the nervous system collect information, the brain and spinal cord integrate and interpret information and form a response, and the motor system carries out the response.

The nervous system is made of nervous tissue. Nervous tissue has two types of cells: neuroglia and neurons. Neuroglia are support cells. There are four in the CNS: astrocytes, microglia, oligodendrocytes, and ependymal cells, and two in the PNS: satellite cells and Schwann cells. Neurons, the nerve cells, carry out the functions of the nervous system: data collection, interpretation, and control of movements.

Neurons are excitable cells. They are capable of acting like batteries, changing their charge when stimulated. The ability to change charge allows neurons to communicate with each other via tiny electrical currents. The electrical activity of neurons can be divided into four basic types of activity. Action potentials are all-or-none events. During an action potential, a neuron first becomes more positive than at rest (depolarization), then returns to rest (repolarization), overshoots and becomes more negative than resting (hyperpolarization), before finally returning to rest. Some neurons are able to depolarize in a graded way, a bigger stimulus causing a bigger depolarization. This kind of electrical activity is called a local potential. Several local potentials can add together to form an action potential. Action potentials, once generated, must flow down the axon of the neuron. The movement of the action potential down the axon is called impulse conduction. The speed of the impulse depends on the diameter of the axon and the presence of myelin around the axon. Disruption of the myelin disrupts information flow around the nervous system. Multiple sclerosis and Guillain-Barré syndrome are disorders caused by destruction of myelin. For a neuron to communicate with another excitable cell, information must move from the neuron to another cell. This information transfer happens at synapses and is called synaptic transmission. The two types of synapses are electrical synapses and chemical synapses. Chemical synapses require the use of neurotransmitters. Electrical synapses do not.

The central nervous system consists of the brain and spinal cord. The spinal cord is an information pathway carrying information between the brain and the body below the neck. The spinal cord is a hollow tube divided in half. It has 31 segments named for the vertebrae. Each segment has a pair of spinal nerves, which carry both sensory and motor information. Covering the spinal cord are three membranes known as meninges: the dura mater, the arachnoid mater, and the pia mater. The inside of the spinal cord consists of central gray matter (cell bodies) surrounded by white matter (myelinated axons). The gray matter is divided into dorsal, lateral, and ventral horns on each side. The white matter is divided into dorsal, lateral, and ventral columns on each side. In the white matter are a series of spinal cord pathways that carry information to and from the brain. Some simple behaviors, known as reflexes, are mediated by the spinal cord alone, without communication with the brain, but complex functions require communication between the brain and spinal cord.

Damage to the columns and/or horns causes sensory and motor problems. Polio destroys the motor neurons in the spinal cord, causing paralysis. Postpolio syndrome strikes long-term polio survivors. Peripheral neuropathy causes sensory and motor loss in the part of the body served by the damaged nerve. Spinal cord injuries may completely stop information flow between the brain and spinal cord, making it impossible for the brain to control or monitor activity below the injury.

The brain is the ultimate controller of the nervous system and therefore of body activities. It consists of several parts. The cerebrum, the largest part of the brain, is divided into five lobes. Four of them are named for skull bones. Each lobe has a unique set of functions. The cerebrum is divided into two hemispheres (halves), and its surface is not smooth but covered in ridges known as convolutions, made up of gyri and sulci. The cerebellum, the little brain, is posterior and inferior to the cerebrum. It resembles the cerebrum. The brain stem is a stalk on which the cerebrum sits, and consists of three parts: the medulla oblongata, the pons, and the midbrain. The brain is also covered by meninges. Damage to the brain either from injury or lack of oxygen disrupts the function of the brain. Symptoms vary depending on the severity and the location of the damage. Mild brain injuries often allow for complete recovery, whereas severe or widespread injuries often result in permanent disability or even death.

Internally, the brain consists, like the spinal cord, of gray and white matter. The cerebrum and cerebellum have central white matter surrounded by a gray matter cortex. There are also deep areas of gray matter surrounded by white matter. Each part of the gray matter cortex is specialized for a particular function. The deep gray matter areas are called nuclei. Each nucleus has a unique function. Deep to the cerebrum is a part of the brain called the diencephalon. The diencephalon contains relay stations between different parts of the CNS and areas that control basic physiology and homeostasis.

Like the spinal cord, the brain is also hollow, with fluid-filled spaces called ventricles. The fluid filling the ventricles and the spinal cord is known as cerebral spinal (CSF) fluid. It is made in the ventricles and carries out many of the functions associated with blood in other parts of the body. There is no blood in the CNS. Excess CSF causes a condition called hydrocephalus. Untreated hydrocephalus can cause permanent damage to the brain.

Attached to the surface of the brain are 12 pairs of cranial nerves. These nerves allow communication between the brain and peripheral structures just as the spinal nerves do for the spinal cord. Cranial nerves may be sensory or motor or both.

The somatic sensory system controls your sense of touch. Information comes into the CNS from the body via spinal and cranial nerves. The information is carried via ascending spinal cord pathways (if from below the neck) to the brain. The information first stops at the thalamus, where it is modified and integrated with other information. Then the information projects to a part of the parietal lobe called the primary somatic sensory cortex in the postcentral gyrus. The postcentral gyrus contains a sensory map of the body. The size of the map for each part of the body is related to the sensitivity of the body part. Once the information reaches the postcentral gyrus, you become aware of the sensation. Understanding of the sensation only occurs when the information is interpreted by the somatic sensory association area in the parietal lobe. Other senses are registered and interpreted in other parts of the brain.

The somatic motor system controls voluntary movements. Movements are planned by the premotor cortex in the frontal lobe and sent to the primary motor cortex in the precentral gyrus of the frontal lobe. The precentral gyrus has a motor map much like the sensory map in the postcentral gyrus. The motor map for each body part is proportional to fine motor control of the body part. The orders are sent from the primary motor cortex to the brain stem and spinal cord via two routes: directly and indirectly. The direct pathways send orders directly to ventral horn and brainstem motor neurons. The indirect pathways, which are used for coordinating movements, project first to subcortical structures like the thalamus, basal nuclei, and the cerebellum. The information passes through a complex coordination loop before projecting to the spinal cord. This information pathway makes sure that the movements are accurate. Disruption of any part of the motor system causes paralysis. Three examples of motor disorders are cerebral palsy (CP), Parkinson's disease (PD), and amyotrophic lateral sclerosis (ALS).

Involuntary movements and physiology are controlled by the autonomic nervous system. The autonomic nervous system is divided into two branches: sympathetic and parasympathetic. The parasympathetic branch controls everyday activities, decreasing respiration, blood pressure, and heart rate and increasing digestion and urination. The sympathetic branch controls flight-or-fight response, increasing heart rate, respiration rate, and blood pressure and decreasing digestion and urination. There are several anatomical and physiological differences between the branches.

Any disruption of any part of the nervous system can have severe consequences. The system is very tightly controlled, and any damage decreases the ability of the nervous system to control body functions. At the end of the chapter is a list of nervous system disorders. The table is only a short list of all the possible disorders of this important control system.

CHAPTER OUTLINE

I. Overview

II. Nervous tissue
   A. Neuroglia: support cells
      1. CNS
      2. PNS
   B. Neurons
      1. Classification by structure
      2. Classification by function

III. How neurons work
   A. Excitable cells
   B. Action potential
   C. Local potential
   D. Impulse conduction
   E. Synapses
      1. Electrical
      2. Chemical
   F. Pathology
      1. Epilepsy
      2. Multiple sclerosis
      3. Guillain-Barré syndrome

IV. Spinal cord and spinal nerves
   A. External anatomy
   B. Meninges
   C. Internal anatomy
      1. Anatomy
      2. Ascending pathways
      3. Descending pathways
   D. Spinal nerves
   E. Reflexes
   F. Pathology
      1. Polio and postpolio syndrome
      2. Peripheral neuropathy
      3. Spinal cord injury

V. Brain and cranial nerves
   A. External anatomy
      1. Cerebrum
      2. Cerebellum
      3. Brain stem
      4. Meninges
   B. Internal anatomy
      1. Cerebrum
      2. Diencephalon
      3. Cerebellum
   C. Ventricles and CSF

D. Cranial nerves
E. Pathology
1. Traumatic brain injury and stroke
2. Alzheimer's disease
3. Hydrocephalus

VI. Big picture: integration
A. Somatic sensory system
B. Somatic motor system
1. Cortical
2. Subcortical
3. Pathology
a. Cerebral palsy
b. Parkinson's disease
c. Amyotrophic lateral sclerosis

VII. Autonomic nervous system
A. Sympathetic
B. Parasympathetic

MEDICAL TERMINOLOGY REVIEW

Define the following terms.

1. Neuropathy: _____

2. Paralysis: _____

3. TBI: _____

4. CVA: _____

5. Persistent unconsciousness: _____

6. Minimally conscious state: _____

7. Concussion: _____

8. Dementia: _____

9. Spinal cord injury: _____

10. Progressive: _____

MULTIPLE CHOICE

Circle the letter of the correct answer.

1. The function of the lateral horn is:
   a. fusing the spinal nerves.
   b. controlling the autonomic nervous system.
   c. communication between multipolar motor neurons and bipolor motor neurons.
   d. acting as a superhighway for myelinated neurons up and down the spinal cord.

2. What is the name of lumbar, sacral, and coccygeal spinal nerves that look like a horse's tail and extend laterally away from a very small area, then route their fibers downward?
   a. Equinal ropa
   b. Cauda equina
   c. Equine threades
   d. Linea alba

3. Both sets of spinal roots fuse to form:
   a. spinal cord.
   b. spinal nerve.
   c. ganglion.
   d. conus medullaris.

4. The dorsal horn of the spinal cord is involved with:
   a. production of cerebrospinal fluid.
   b. motor function.
   c. sensory function.
   d. coordination.

5. Choose the correct order of the CNS's protective membrane from innermost to outermost layer.
   a. Pia mater, dura mater, arachnoid
   b. Archnoid, dura mater, pia mater
   c. Pia mater, arachoid, dura mater
   d. Dura mater, arachnoid, pia mater

6. How many nerves enter and exit at the cervical region?
   a. 31
   b. 6
   c. 24
   d. 8

7. In the CNS, the glial cells that cover and line cavities are:
   a. ependymal.
   b. oligodendrocytes.
   c. Schwann cells.
   d. astrocytes.

8. The nervous system has an output side called:
   a. motor.
   b. sensory.
   c. parasympathetic.
   d. neuroglia.

9. For repolarization to occur, which of the following is true about movement of ions?
   a. Potassium moves out of cells.
   b. Calcium moves into cells.
   c. Sodium moves into cells.
   d. Chlorine moves out of cells.

10. Between adjacent Schwann cells are bare spots where channels must open for action potentials to flow down the axon with haste. What are they called?
    a. Nodes of Ranvier
    b. Conus medullaris
    c. Myelin
    d. Spudus impulsasis

11. Which of the following axon characteristics will constitute the slowest ionic flow?
    a. Wide, myelinated
    b. Narrow, unmyelinated
    c. Wide, unmyelinated
    d. Narrow, myelinated

12. Which of the following branches in the body's alert system is commonly known as "fight or flight"?
    a. Sympathetic
    b. Parasympathetic
    c. Central nervous system
    d. Both a and b

13. Multipolar neurons have:
    a. many cell bodies; single dendrite, single axon.
    b. unlimited ionic transferability.
    c. many dendrites, single axon.
    d. one dendrite, multiple axons.

14. The area of the cerebral cortex that allows understanding and interpretation of somatic sensory information is called the:
    a. somatic sensory association area.
    b. precentral gyrus.
    c. diencephalon.
    d. limbic reticular formation.

15. The map size in the motor cortex is proportional to:
    a. the amount of movement control.
    b. the size of the structure.
    c. the blood supply.
    d. the insula.

16. If you violently stub your toe (pain), which of the following pathways carries the information to the brain?
    a. Dorsal column tract
    b. Spinothalamic tract
    c. Spinocerebellar tract
    d. Spinobulbar tract

17. The diencephalon is made up of which of the following structures?
    a. Pineal, hypothalamus, pituitary gland, and thalamus
    b. Midbrain, cerebellum, pituitary, and corpus callosum
    c. Midbrain, thalamus, hypothalamus, and cerebellum
    d. Pineal, pituitary, adrenal gland, and midbrain

18. The parietal lobes are mainly responsible for:
    a. motor activities.
    b. sensory perception.
    c. vision.
    d. integration of emotions.

19. Where is cerebrospinal fluid made?
    a. In the subarachnoid and subdural spaces and central canal by the pituitary glands
    b. In the fourth ventricle by the cerebral aqueduct
    c. In the lateral ventricle(s) by the choroid plexus
    d. In the hypothalamus by the thalamal peptides

20. The function of the cerebellum is:
    a. reflex center for cough and sneeze.
    b. motor coordination and balance.
    c. coordination of heart rate with breathing rate.
    d. interpretation of crude touch and temperature.

21. The function of the midbrain is:
    a. coordination.
    b. two-way conduction system pathway to relay visual and auditory impulses.
    c. wisdom, moderation of impulse, and conscience.
    d. production of cerebral spinal fluid.

22. Ventricles of the brain are:
    a. networks of associative neurons that link all four named lobes to each other.
    b. hollows or spaces in the brain for sound resonance and to make the brain lighter.
    c. ridges and crevices on the surface of the brain.
    d. fluid-filled cavities.

23. Where is the third ventricle?
    a. Between the cerebrum and the brain stem
    b. In the cerebrum between the frontal and parietal lobes
    c. In the cerebrum spanning the temporal, frontal, and parietal lobes
    d. Between the cerebellum and the medulla

24. The smell of citrus causes Mary to feel anxious. When Mary was a child, her grandmother, who was verbally abusive, served fresh orange juice in the morning before her predictable morning tantrum. What system of the brain coordinates emotion and sense of smell as well as retrieves memories?
    a. Limbic
    b. Sympathetic
    c. Parasympathetic
    d. Somatic sensory

25. Where are the sympathetic preganglionic neurons located?
    a. In the thoracic and lumbar segments of the spinal cord
    b. In the cranial and sacral segments of the spinal cord
    c. Close to the visceral or glandular organs they affect
    d. Running parallel to the vertebral column/spinal cord

26. _____, a peripheral demyelinating disease, is characterized by sudden onset, ascending paralysis.
    a. Multiple sclerosis
    b. Guillain-Barré syndrome
    c. Charcot-Marie-Tooth disorder
    d. Myasthenia gravis

27. Polio has been eradicated in North America. There have been no cases of polio in the United States for nearly 30 years. Why is polio still a health issue in the United States?
    a. Foreign travel
    b. The virus is hiding somewhere
    c. Postpolio syndrome
    d. It's not an issue

28. Damage to peripheral nerves, from many different causes, is known as:
    a. peripheral neuropathy.
    b. peripheral nerve syndrome.
    c. peripheral movement disorder.
    d. peripheral myopathy.

29. A mild traumatic brain injury is known as a(n) _____, whereas a severe traumatic brain injury may leave patients in a(n) _____, a potentially permanent condition.
    a. stupor; concussion
    b. persistent unconsciousness; stupor
    c. minimally conscious state; persistent unconsciousness
    d. concussion; minimally conscious state

30. What are the four As of Alzheimer's symptoms?
    a. Anger, aggression, anxiety, apathy
    b. Anger, appetite, aggression, apathy
    c. Anger, appetite, agility, anxiety
    d. Anger, aggression, agility, apathy

31. Which of the following is diagnostic for Parkinson's disease?
    a. It's congenital
    b. Foot drag
    c. Cogwheel rigidity
    d. Spastic paralysis

32. This disease is one of the only disorders with both flaccid and spastic paralysis.
    a. Cerebral palsy
    b. Amyotrophic lateral sclerosis
    c. Parkinson's disease
    d. Alzheimer's disease

33. Post-injury, there is great danger of increased CNS damage due to:
    a. increased pressure.
    b. release of neurotransmitters.
    c. inflammation.
    d. All of the above

34. After a skiing accident, a young man presents to the emergency department with bilateral paralysis. He is conscious but unable to breathe on his own. Where is his injury?
    a. T1
    b. C7
    c. Parietal lobe
    d. C1

35. Jim has been in poor health for years. He has periods of confusion and anger. His movements have become slow, and he shakes when not moving. What is wrong with Jim?
    a. Huntington's disease
    b. Parkinson's disease
    c. Alzheimer's disease
    d. Guillain-Barré syndrome

MATCHING EXERCISES

Set 1

Please match each term with the appropriate definition.

_____ 1. Neuron
_____ 2. Astrocytes
_____ 3. Satellite cells
_____ 4. Dura mater
_____ 5. Microglia
_____ 6. Arachnoid mater
_____ 7. Schwann cells
_____ 8. Oligodendrocyte
_____ 9. Pia mater
_____ 10. Ependymal

a. Precursor of the neuroglial cells
b. Metabolic and structural support cells of CNS
c. Support cells of PNS
d. Nerve cells for functional control of the nervous system
e. Covers and lines cavities of the CNS
f. Innermost meningeal layer
g. Outermost meningeal layer
h. Middle meningeal layer
i. Cells that remove debris from the CNS
j. Makes myelin for the CNS
k. Makes myelin for the PNS

Set 2

Please match each brain part with the appropriate function.

_____ 1. Medulla oblongata
_____ 2. Parietal lobe
_____ 3. Hypothalamus
_____ 4. Pineal gland
_____ 5. Frontal lobe
_____ 6. Cerebellum
_____ 7. Corpus callosum
_____ 8. Temporal lobe
_____ 9. Pons
_____ 10. Occipital lobe

a. Secretes epinephrine
b. Connect left and right hemispheres
c. Primary area for hearing function
d. Primary area for motor function
e. Divides the brain into hemispheres
f. Primary area for vision functions
g. Coordination
h. Relays sensory and motor information
i. Regulates heart rate, blood pressure, vomiting
j. Regulate body temperature, fear, and pleasure
k. Secretes melatonin
l. Primary area for sensory functions

Set 3

Please match each cranial nerve with the appropriate function.

_____	1. Facial nerve	a. Vision
_____	2. Oculomotor nerve	b. Movement of the eye, C. N. IV
_____	3. Glossopharyngeal nerve	c. Sensation of the face and muscles for chewing
_____	4. Accessory nerve	d. Muscles of expression like squinting and smiling
_____	5. Olfactory nerve	
_____	6. Trochlear nerve	e. Smell
_____	7. Optic nerve	f. Swallowing and taste, C. N. IX
_____	8. Hypoglossal nerve	g. Movement of the tongue
_____	9. Trigeminal nerve	h. Movement of the shoulder's trapezius muscles
_____	10. Vagus nerve	i. Movement of the eye, C. N. III
		j. Sensation of the stomach (bellyache)
		k. Hearing and balance

Set 4

Please match each disorder with the appropriate symptoms.

_____	1. Guillain-Barré syndrome	a. Unilateral motor/sensory loss
_____	2. Multiple sclerosis	b. Personality change, memory loss, confusion
_____	3. Peripheral neuropathy	c. Local motor and sensory malfunction
_____	4. Hydrocephalus	d. Nonprogressive motor deficits
_____	5. Cerebral palsy	e. Progressive paralysis, normal sensation
_____	6. Huntington's disease	f. Rapid onset, ascending paralysis
_____	7. Amyotrophic lateral sclerosis	g. Chorea, mood swings, memory loss
_____	8. Parkinson's disease	h. Skull expansion, seizures, dementia
_____	9. Alzheimer's disease	i. Shuffling gate, resting tremor, cogwheel rigidity
_____	10. Cerebrovascular accident	j. Symptoms vary

FILL IN THE BLANK

Fill in the blanks to complete the following statements.

1. The parasympathetic division of the nervous system is often called the
   _____ system.

2. The combination of axon terminal and receiving neural dendrite is called
   a _____.

3. The vesicles at the axon terminal are filled with chemicals called
   _____.

4. The three horns of the spinal cord are the
   _____,
   the _____, and the
   _____ horns.

5. The _____ root is sensory, whereas the

   _____ root is motor.

6. The simplest form of motor output that protects us from harm is a(n)

   _____.

7. The three-layered protective membrane of both the spinal cord and brain is called the _____.

8. When a nerve cell is stimulated,

   _____ ions rush into

   the cell.

9. The surface of the cerebrum has broken ridges called

   _____.

10. The occipital lobes are responsible for

    _____.

11. The three sections of the brain stem are the

    _____, the

    _____, and the

    _____.

12. The layer of gray matter surrounding the white matter of the brain is called _____.

13. The ventricles of the brain contain

    _____.

14. A remarkable woman named Harriet Tubman was born into slavery but had the courage to free herself, hundreds of kidnapped Africans, and their descendants. Ironically, considering that she was constantly and dangerously on the run, she suffered from narcolepsy, which means she would fall asleep uncontrollably. As a child, due to a violent blow by the plantation overseer, Harriet Tubman's

    _____ of the brain

    was damaged. This part of the brain is vital in the maintenance of conscious awareness.

15. The corticospinal and corticobulbar tracts carry

    _____ signals to synaptic

    junctions in the ventral horn of the spinal cord.

16. Autoimmune destruction of

    _____ in the CNS is a

    disorder called multiple sclerosis.

17. _____ is caused by a viral

    infection that destroys ventral horn motor neurons.

18. The most common cause of spinal cord injuries is

    _____.

19. If a spinal cord injury occurs above

    _____, the patient will

    often be ventilator dependent.

20. A lumbar puncture is also known as a(n)

_____.

21. Blockage or narrowing of passages between ventricles leads to a condition
known as _____, which
literally means "water in the head."

22. _____ is a congenital
paralysis with many different causes.

23. Most patients with amyotrophic lateral sclerosis (ALS) die from

_____.

24. _____ is a genetic disease
causing movement problems and dementia.

25. Mabel, a 75-year-old woman with a history of transient ischemic attacks,
awakens one morning unable to speak or move her right side. She has
had a major stroke affecting the _____
_____ side of her cerebral cortex.

SHORT ANSWER

1. How does the size of a stimulus affect the resulting local potential?

_____

_____

_____

_____

2. What is the primary difference between action potentials and local
potentials?

_____

_____

_____

_____

3. What is meant by *contralateral* information entering and leaving the brain?

_____

_____

_____

_____

4. Besides location, what are two differences between the spinal nerves and the cranial nerves?

   _____

   _____

   _____

   _____

5. What effects do both the sympathetic and parasympathetic nervous systems have on skeletal muscle, cardiac muscle, and the muscle surrounding the digestive tract?

   _____

   _____

   _____

   _____

6. Why is it so difficult to diagnose nervous system disorders?

   _____

   _____

   _____

   _____

7. Explain the difference between flaccid and spastic paralysis.

   _____

   _____

   _____

   _____

LABELING ACTIVITIES

1. Label the parts of the spinal cord using Figure 9-9 in your textbook as a guide.

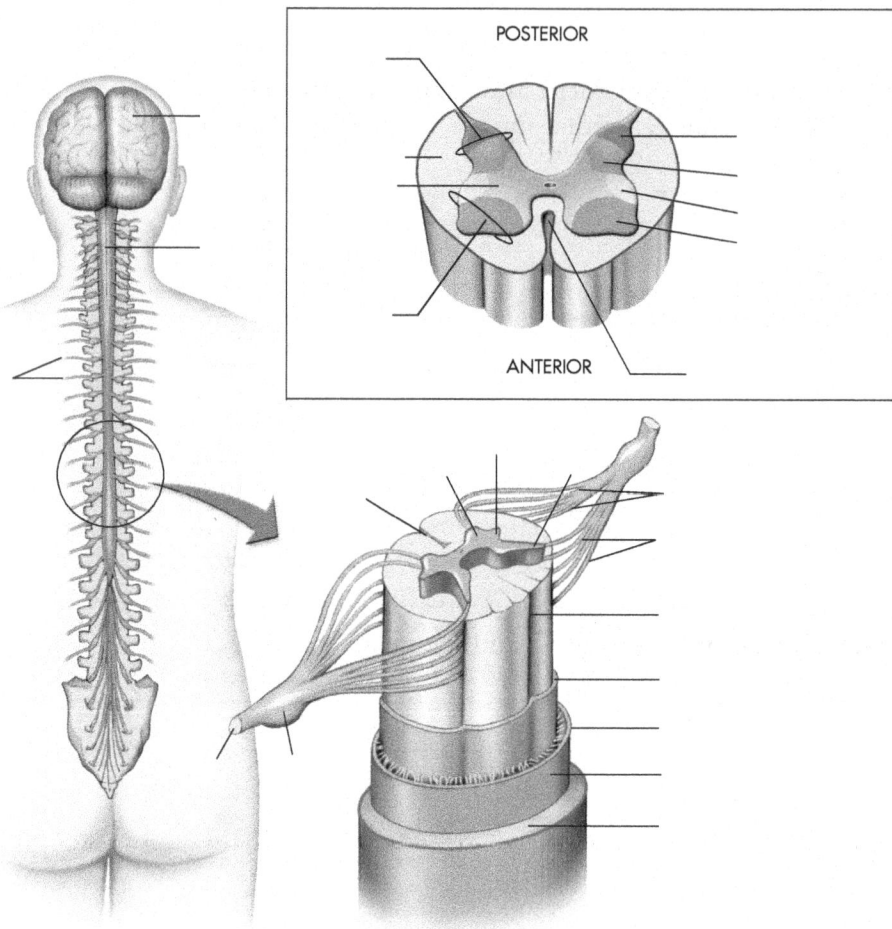

POSTERIOR

ANTERIOR

2. Identify the following structures. Then color code the following brain regions: cerebral cortex, interior cerebrum, diencephalon, midbrain, pons, medulla oblongata, using Figure 9–15 in your textbook as a guide.

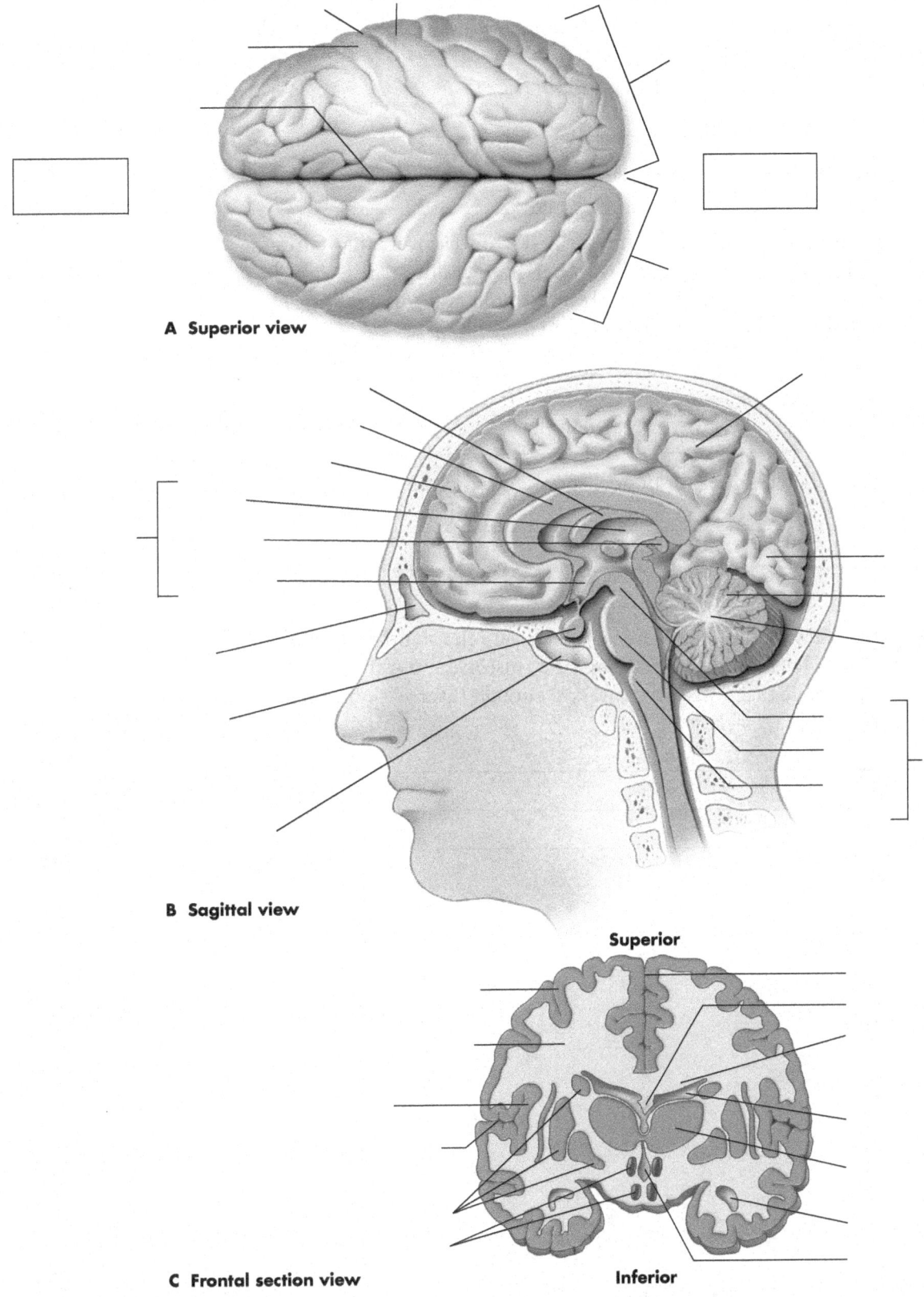

A Superior view

B Sagittal view

C Frontal section view

Superior

Inferior

CASE STUDY

Ginny is a 65-year-old woman who had been healthy until recently. In the last month or so she has become increasingly confused, has trouble concentrating, and gets dizzy and anxious. She often complains of headaches. Her worried daughter forces Ginny to see her primary care physician, who refers her to a neurologist.

1. What is the first thing the neurologist should do?

   _____

   _____

   _____

2. After talking to Ginny and her daughter, the doctor discovers that Ginny recently fell and hit her head playing with her grandchildren. She was not knocked unconscious but had an obvious bump and some passing dizziness. What tests should be run on Ginny?

   _____

   _____

   _____

3. What are the possible causes for her collection of symptoms?

   _____

   _____

   _____

4. Ginny's MRI, x-ray, and CT scan are normal, as are the standard blood and urine tests. Given Ginny's age, symptoms, history, and test results, what is the probable diagnosis? Do you have enough information to tell what is wrong at this stage?

   _____

   _____

   _____

LEARNING ACTIVITIES

1. For each brain region, list the deficits that would result after CVA or TBI.

2. Stem cells have been promoted as a potential cure for degenerative diseases. Pick one progressive neurological disease, and research the role of stem cells as a potential treatment. How close are scientists to using stem cells to treat these disorders? Have each student in your group pick a different disease and explain the role of stem cells in treatment.

3. Play "Neuro Concentration." Make a deck of cards. On one card print a nervous system structure. On the matching card print the function. Turn the cards over, keeping the ones that match.

4. Many well-known individuals, especially athletes, have been victims of neurological disorders or injuries. Do some research using the Internet or news sources. What happens to a person after diagnosis of a neurological disorder, even if they have the best treatment? How do they cope? What happens to them? Some examples: Christopher Reeve, Muhammad Ali, Lou Gehrig, Andy Griffith, Ronald Reagan, and Michael J. Fox.

5. In 1982 several heroin addicts in California were poisoned by a chemical called MPTP and got instant, profound Parkinson's. Read their story. How did they develop Parkinson's? How did their illness increase our understanding of Parkinson's? (Hint: Search for "frozen addicts.")

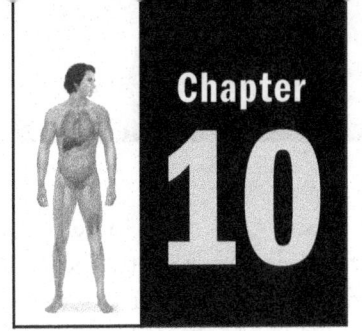

THE ENDOCRINE SYSTEM: THE BODY'S OTHER CONTROL CENTER

 ## CHAPTER SUMMARY

Your body's second control system is the endocrine system, a series of glands and organs that secrete chemical messengers called hormones into the bloodstream. Hormones are long-lasting chemical messengers that affect the metabolic activities of cells. The target cells are often far away from the endocrine organ that secretes the hormone. Hormones affect cells in two ways. Hormones may bind to extracellular receptors, causing permeability or enzyme changes in the target cell. In addition, special hormones called steroids cross the cell membrane and bind to intracellular receptors inside the cell. Steroids interact directly with DNA, making them extremely potent hormones. Hormone levels in the body are tightly controlled, mainly by negative feedback. There are three sources of control of hormone levels: control by another endocrine organ, control by the nervous system, and humoral control.

There are several endocrine glands. The hypothalamus, part of the diencephalon of the brain, is the major link between the nervous and endocrine systems because the hypothalamus is part of both. The hypothalamus makes and secretes many different hormones, most of which control the activity of another endocrine gland, the pituitary. The pituitary gland has two parts: the posterior pituitary, which is an extension of the hypothalamus, and the anterior pituitary, which makes and secretes its own hormones. The hypothalamus makes two hormones: oxytocin and antidiuretic hormone, both of which are secreted by the posterior pituitary. The rest of the hypothalamic hormones are inhibiting or releasing hormones that control secretion of anterior pituitary hormones. The hormones from the anterior pituitary control tissue metabolism and the activity of other endocrine organs. The fact that the anterior pituitary controls other endocrine organs has earned it the name "master gland." Because pituitary hormones affect so many different tissues, pituitary disorders have widespread consequences. For example, Cushing's syndrome, a disorder in which the anterior pituitary overproduces hormones, causes high blood cholesterol, high blood sugar, obesity, and depression.

The thyroid gland, in the anterior neck, produces three hormones. Two of the hormones, T3 and T4, contain iodine and control cellular metabolism. Too little hormone (hypothyroidism) causes fatigue, mental confusion, and weight gain. Too much hormone (hyperthyroidism) causes anxiety, tachycardia, and weight loss. Autoimmune disorders are the most common cause of thyroid malfunction. Graves' disease causes hyperthyroidism, and Hashimoto's thyroiditis causes hypothyroidism. The third thyroid hormone is calcitonin. Calcitonin decreases blood calcium. Blood calcium is increased by parathyroid hormone. The parathyroid glands are small glands on the posterior surface of the thyroid.

The thymus gland, in the thoracic cavity, secretes thymosin, which stimulates the maturation of white blood cells. It is important in children. The pineal gland, part of the diencephalon, secretes melatonin, which controls our sleep-wake cycles.

The pancreas is responsible for controlling blood sugar using two hormones: insulin and glucagon. After a meal, as blood glucose rises, the pancreas secretes insulin. Insulin enables glucose to get into cells and directs the liver to store excess glucose as glycogen. Thus, insulin decreases blood glucose. Later after the meal, as blood glucose falls, the pancreas secretes glucagon. Glucagon directs the liver to break glycogen down into glucose and release it into the bloodstream, increasing blood glucose. Diabetes mellitus is a disorder that results from auto-immune destruction of insulin producing cells (Type 1 or early onset) or from the insensitivity of body cells to insulin (Type 2 or late onset). In either case, the most obvious sign of diabetes is excessively high blood glucose. Without the effects of insulin, glucose cannot get into cells or be stored by the liver. When blood glucose gets too high, kidneys work overtime to try to get rid of the excess glucose, causing increased urination and thirst. In addition, cell metabolism is abnormal due to lack of glucose inside cells, causing abnormalities in blood chemistry.

The adrenal glands, on top of the kidneys, can be divided into two parts: the adrenal medulla and the adrenal cortex. The adrenal medulla is an extension of the sympathetic nervous system. When the sympathetic nervous system releases epinephrine, the adrenal medulla is stimulated to release epinephrine and norepinephrine as circulating hormones. These hormones have the same effects as the sympathetic nervous system. The adrenal cortex secretes several steroid hormones. These hormones are involved in nearly every physiological process in the body. Mineralcorticoids, for example, control electrolyte balance, whereas glucocorticoid hormones control blood sugar. The steroid hormones are tightly controlled. Steroids, whether prescribed or used illegally, can cause many side effects. Cortisol, one of the adrenal cortex hormones, is involved in stress response. The initial response to stress is appropriate and even helpful. If stress becomes chronic, however, the effects are detrimental. The table at the end of this chapter has a list of several endocrine disorders.

 CHAPTER OUTLINE

  I. Organization of endocrine system
      A. Endocrine organs
      B. Hormones

  II. Control of endocrine activity
      A. Negative feedback
      B. Sources of control

III. Endocrine organs
- A. Hypothalamus
- B. Pituitary
  1. Posterior
  2. Anterior
  3. Pathology
     - a. Diabetes insipidus
     - b. Syndrome of Inappropriate ADH
     - c. Hypopituitarism
     - d. Hyperpituitarism
     - e. Stature disorders
- C. Thyroid gland
  1. Pathology
     - a. Hypothyroidism/Hashimoto's thyroiditis
     - b. Hyperthyroidism/Graves' disease
- D. Parathyroid gland
- E. Thymus gland
- F. Pineal gland
- G. Pancreas
  1. Pathology: diabetes mellitus
- H. Adrenal glands
  1. Adrenal medulla
  2. Adrenal cortex
  3. Pathology
     - a. Addison's disease
     - b. Cushing's syndrome
     - c. Side effects of steroids
     - d. Steroid abuse
     - e. Cortisol and stress response
- I. Gonads

IV. List of endocrine disorders

MEDICAL TERMINOLOGY REVIEW

Define the following terms.

1. Hypothyroidism: _____

2. Hyperthyroidism: _____

3. Hormone replacement therapy: _____

4. Negative feedback: _____

5. Endocrine: _____

6. Hormone: _____

7. Steroids: _____

8. Acromegaly: _____

9. Diabetes mellitus: _____

10. Goiter: _____

MULTIPLE CHOICE

Circle the letter of the correct answer.

1. What is the target organ(s) for glucagon?
   a. Pancreas
   b. Kidneys
   c. Adrenals
   d. Liver

2. How do hormones and neurotransmitters (NT) differ?
   a. Hormones are secreted by exocrine glands, and NTs are secreted from endocrine glands.
   b. Hormones are fast to take action, and NTs are slow to take effect.
   c. Hormones are secreted by endocrine glands, and NTs are secreted from axon terminals.
   d. Both b and c

3. Where are the adrenal glands located?
   a. Above the kidneys
   b. In the brain stem
   c. In the chest
   d. In the neck

4. How is insulin controlled?
   a. Positive feedback
   b. Negative feedback
   c. Neutral feedback
   d. Neural feedback

5. Which hormone needs iodine for production?
   a. Insulin
   b. Thymosin
   c. Iodomelanoin
   d. Thyroxine

6. Which gland or organ secretes releasing and inhibiting hormones controlling the master gland?
   a. Pituitary
   b. Adrenal
   c. Pancreas
   d. Hypothalamus

7. Where is the pancreas located?
   a. In the abdomen
   b. In the brain
   c. In the neck
   d. In the pelvis

8. What directly influences the production of testosterone?
   a. Gonadotropin-releasing hormone
   b. Corticotropic hormone
   c. Luteinizing hormone
   d. ACTH

9. What gland is at its greatest size and efficiency in childhood, fighting infection and helping in the maturation of white blood cells?
   a. Thyroid
   b. Thymus
   c. Testis
   d. Adrenals

10. Which of the following is a function of one of the hormones secreted by the adrenal cortex?
    a. Fight-or-flight
    b. Salt and fluid balance
    c. Skin pigmentation
    d. Iodine production

11. What is the target organ for ACTH?
    a. Adrenal glands
    b. Adenoids
    c. Anterior pituitary
    d. Arterial walls

12. What is true about hormones?
    a. They focus on targets very close to the gland that secretes them.
    b. They affect a single cell.
    c. Their effects wear off quickly.
    d. Their effects are long lasting.

13. Why are steroid hormones so powerful?
    a. Steroid hormones pass easily through the target cell membrane and interact with the cell's DNA.
    b. They are always secreted in great amounts.
    c. They interact with the neuronal cell bodies as well as the neural membranes.
    d. All of the above

14. What is the target organ for ADH?
    a. Adenoids
    b. Kidneys
    c. Adrenals
    d. Pancreas

15. Which hormone antagonizes glucagon?
    a. Insulin
    b. Glycogen
    c. Thymosin
    d. Calcitonin

16. Where is the pineal gland located?
    a. In the chest
    b. In the pelvis
    c. In the neck
    d. In the brain

17. What may occur if calcitonin is hypersecreted?
    a. Low blood calcium
    b. High blood cholesterol
    c. Low blood sugar
    d. High blood viscosity

18. Where is melanocyte-stimulating hormone produced?
    a. Posterior pituitary
    b. Thyroid
    c. Parathyroid
    d. Anterior pituitary

19. Which gland is located in the chest?
    a. Thyroid
    b. Thymus
    c. Pineal
    d. Pituitary

20. The diencephalon is home to the:
    a. hypothalamus.
    b. adrenals.
    c. thymus.
    d. pancreas.

21. Which hormone increases the release of LH and FSH from their releasing gland?
    a. Adrenocorticotropic hormone
    b. Estrogen
    c. Adrenocorticosteroids
    d. Gonadotropin-releasing hormone

22. Parathyroid hormone targets the:
    a. bladder.
    b. skin.
    c. bones.
    d. mammary glands (breasts).

23. Polyuria (increased urination) is a symptom of:
    a. Addison's disease.
    b. diabetes mellitus.
    c. Cushing's syndrome.
    d. Hashimoto's thyroiditis.

24. For uterine contraction during childbirth the expectant mother needs to secrete:
    a. iodine.
    b. estrogen.
    c. oxytocin.
    d. prolactin.

25. A tumor of the adrenal gland resulting in excess secretion of epinephrine is:
    a. Hashimoto's thyroiditis.
    b. pheochromocytoma.
    c. Addison's disease.
    d. diabetes mellitus.

26. Too much alcohol can make you feel pretty miserable the next day because it:
    a. decreases urination.
    b. acts on receptors to increase urination.
    c. causes diabetes.
    d. All of the above

27. Jenny has recently gained a lot of weight. Her blood pressure and blood sugar are much higher than normal for her, and she has been depressed and sleepy. What might be wrong with Jenny?
    a. Hyperthyroidism
    b. Cushing's syndrome
    c. Acromegaly
    d. Addison's disease

28. Symptoms of _____ include feeling hot, anxiety, irritability, tremors, muscle weakness, and tachycardia.
    a. Cushing's syndrome
    b. hypothyroidism
    c. hyperthyroidism
    d. Addison's disease

29. In Graves' disease, the immune system stimulates TSH receptors causing the thyroid to:
    a. overproduce hormones.
    b. stop producing hormones.
    c. do nothing different.
    d. die.

30. Jim has been diabetic for several years and has managed his disorder well. This morning after his workout he had an episode that really frightened his buddies. In the locker room he was restless and anxious. He seemed confused and irritable. He told his buddies he was fine. Within a few minutes he had collapsed, still conscious, but too weak to stand. His buddies called 911 and were told to do what?
   a. Give him an insulin shot.
   b. Get him a drink of water.
   c. Get some orange juice, soda, or hard candy into him if possible.
   d. Do nothing.

31. Symptoms of Addison's disease include:
   a. darkening skin, weight loss, hypoglycemia, weakness, and low blood pressure.
   b. darkening skin, weight gain, hypoglycemia, weakness, and high blood pressure.
   c. darkening skin, weight loss, hyperglycemia, weakness, and low blood pressure.
   d. darkening skin, weight gain, hyperglycemia, weakness, and high blood pressure.

32. Why are therapeutic steroids prescribed so carefully?
   a. They have serious side effects.
   b. Coming off the drugs requires careful control.
   c. They are banned by both amateur and professional sports organizations.
   d. All of the above

33. Symptoms of chronic stress due to increased cortisol secretion may include:
   a. decreased appetite.
   b. improved immune response.
   c. hypercholesterolemia.
   d. feelings of well-being.

34. The chief treatment for _____ is growth hormone injections.
   a. diabetes mellitus
   b. gigantism
   c. dwarfism
   d. hypothyroidism

35. The most common cause of Cushing's syndrome is:
   a. benign pituitary tumor.
   b. brain tumor.
   c. thyroid tumor.
   d. pancreas tumor.

 MATCHING EXERCISES

Set 1

Please match each gland with the hormone it makes.

_____	1.	Pineal	a.	Antidiuretic hormone
_____	2.	Anterior pituitary	b.	Thyroxine
_____	3.	Posterior pituitary	c.	Melanocyte stimulating hormone
_____	4.	Ovaries	d.	Progesterone
_____	5.	Pancreas	e.	Testosterone
_____	6.	Thymus	f.	Melatonin
_____	7.	Testis	g.	Insulin
_____	8.	Adrenal medulla	h.	Thymosin
_____	9.	Hypothalamus	i.	None
_____	10.	Thyroid	j.	Epinephrine

Set 2

Please match each hormone with the appropriate effect.

_____	1.	Parathyroid hormone
_____	2.	Luteinizing hormone
_____	3.	Follicle-stimulating hormone
_____	4.	Triiodothyronine
_____	5.	Oxytocin
_____	6.	Calcitonin
_____	7.	Insulin
_____	8.	Norepinephrine
_____	9.	Melatonin
_____	10.	Adrenocorticosteroids

a. Increases blood glucose
b. Ovulation
c. Decreases blood calcium
d. Decreases blood sugar
e. Regulates secondary sexual characteristics
f. Increases blood calcium
g. Increases metabolism; secreted by gland located in neck
h. Triggers sleep
i. Milk ejection
j. Prolongs fight-or-flight response
k. Regulates sperm and egg production

Set 3

Please match each disorder with the appropriate signs, symptoms, or cause.

_____	1.	Cushing's syndrome
_____	2.	Addison's disease
_____	3.	Diabetes mellitus
_____	4.	Dwarfism
_____	5.	Bone deterioration
_____	6.	Testicular shrinkage
_____	7.	Graves' disease
_____	8.	Decreased milk production
_____	9.	Impaired ovulation
_____	10.	Hashimoto's thyroiditis

a. Steroid abuse
b. Upper body obesity; high blood sugar
c. Decrease in insulin production or recognition
d. Hyposecretion of prolactin
e. Swollen thyroid gland; hypothyroidism
f. Decreased luteinizing hormone
g. Decreased growth hormone during childhood
h. Hypersecretion of parathyroid hormone
i. Weight loss; low BP; insufficient cortisol
j. Hyperthyroidism; bulging eyes

Set 4

Please match each disorder with the appropriate treatment.

_____	1.	Cushing's syndrome
_____	2.	Addison's disease
_____	3.	Diabetes mellitus
_____	4.	Dwarfism
_____	5.	Diabetes insipidus
_____	6.	Gigantism
_____	7.	Graves' disease
_____	8.	Chronic stress
_____	9.	Hypoparathyroidism
_____	10.	Hashimoto's thyroiditis

a. Calcium, vitamin D, hormone replacement
b. Hormone replacement, drink lots of water
c. Remove pituitary, hormone replacement
d. Growth hormone injections
e. Destroy thyroid gland, hormone replacement
f. Thyroid hormone replacement
g. Diet, exercise, medication, insulin injections
h. Replace adrenocorticosteroid hormones
i. Decrease stress
j. Tumor removal, hormone replacement

FILL IN THE BLANK

Fill in the blanks to complete the following statements.

1. In females, _____
   stimulates the smooth muscle tissue in the wall of the uterus, promoting
   labor and delivery.

2. Another name for the anterior pituitary gland is
   _____.

3. The _____ is embedded in
   the mediastinum, posterior to the sternum.

4. The pancreas secretes
   _____ and
   _____.

5. A hyposecretion of cortisol causes
   _____.

6. T3 and T4 secretions are controlled by
   _____ secreted by the
   anterior pituitary.

7. Milk production is controlled by the hormone
   _____, and milk ejection is
   controlled by a hormone
   _____.

8. The hypothalamus regulates the release of hormones from the
   _____ gland.

9. MSH from the pituitary gland targets
   _____.

10. Alcohol inhibits the hormone
    _____.

11. The three ways hormone levels are regulated are
    _____,
    _____, and
    _____.

12. Prednisone mimics the hormones secreted by the
    _____.

13. The antagonist hormone to calcitonin is secreted by the
    _____ gland.

14. _____ are chemical
    messengers that are released in one tissue and transported by the
    bloodstream to affect target cells and other tissues.

15. The hormone unique to females is
    _____.

16. _____ is a disorder caused
    by the secretion of too little ADH.

17. Patients with traumatic brain injury often have untreated
_____ as late as one year
postinjury, even when they had recovered enough to be released from
their rehabilitation program.

18. Overproduction of _____
may be involved with the pathological consequences of chronic stress.

19. Adults who secrete too much growth hormone have
_____, whereas children
who secrete too much growth hormone have

_____.

20. The most common causes of
_____ is Hashimoto's
thyroiditis.

21. An enlargement of the thyroid called a(n)
_____ can be caused by
either hyperthyroidism or hypothyroidism.

22. Type 1 diabetes mellitus is caused by autoimmune destruction of the
_____ of the pancreas.

23. _____ is a hormone
released by the adrenal cortex in response to stress.

24. _____ is a commonly
prescribed therapeutic steroid.

25. _____ steroids are
commonly abused by athletes, particularly body builders.

SHORT ANSWER

1. Explain the functional difference between an exocrine gland and an
endocrine gland.

_____

_____

_____

_____

2. What are two side effects or risk factors with anabolic steroid abuse that
occur in both men and women?

_____

_____

_____

_____

3. Why does alcohol increase urine output?

_____

_____

_____

_____

4. What is meant by *hormonal* control of hormone levels?

_____

_____

_____

_____

5. How do steroid hormones affect cells?

_____

_____

_____

_____

6. List and briefly explain the three phases of stress response.

_____

_____

_____

_____

7. Several endocrine disorders are autoimmune disorders. List and briefly describe these disorders.

_____

_____

_____

_____

LABELING ACTIVITY

Name the pictured glands using Figure 10–1 in your textbook as a guide.

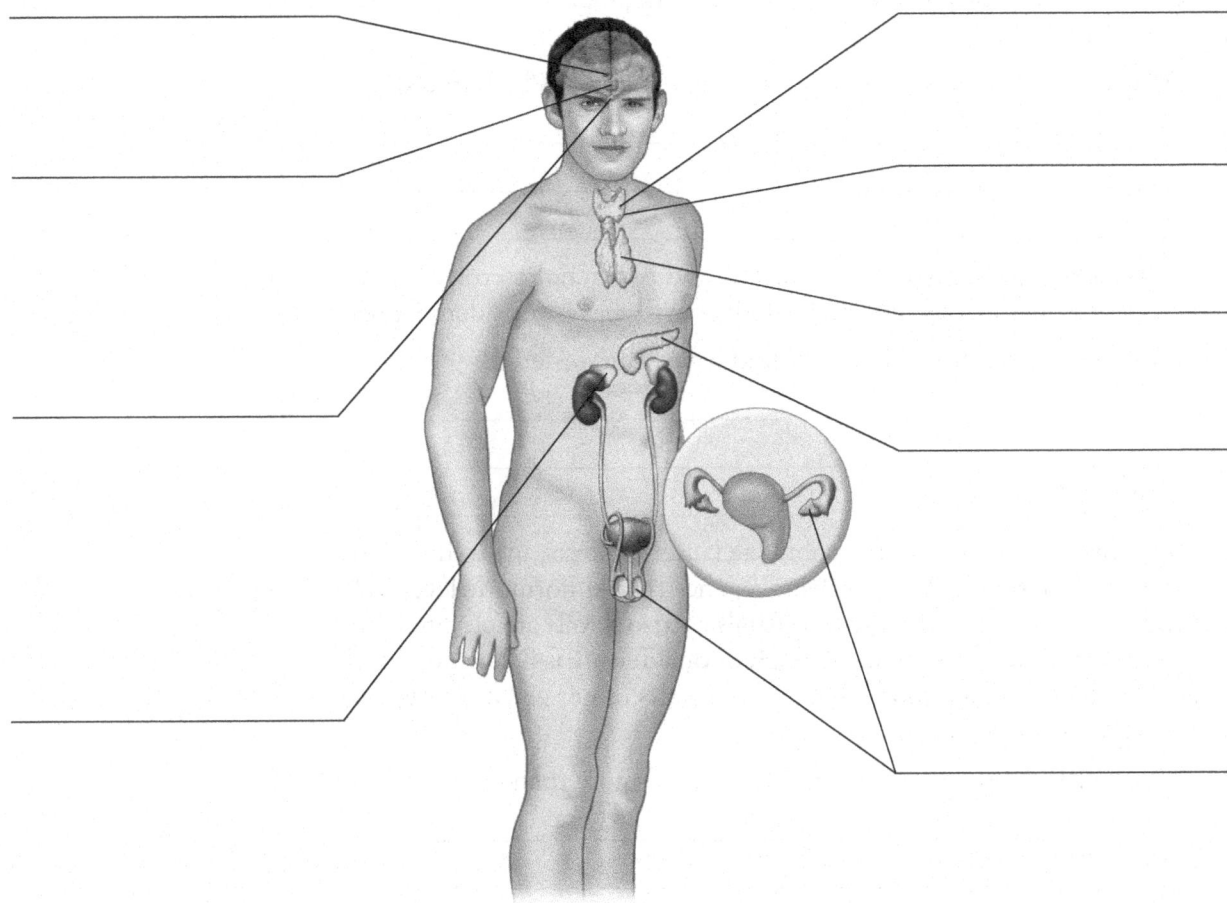

CASE STUDY

Rita, a 50-ish emergency department nurse, has felt pretty awful for several weeks. Overweight and admittedly too sedentary, she has gained a good bit of weight since she quit smoking three years ago. Lately she has been sluggish and depressed, just not herself. Concerned, she has scheduled an appointment with her primary physician.

1. Do Rita's symptoms suggest an endocrine disorder? Which one(s)?

_____

_____

_____

Blood tests show that Rita has hyperglycemia and hypercholesterolemia. She is hypertensive and has a body mass index well above what is considered healthy.

2. What other tests should be, or should have been, ordered?

_____

_____

_____

Rita's physician decides to run the complete battery of tests, including cortisol levels, fasting cholesterol, glucose tolerance, and thyroid hormone levels. He even sends Rita for a bone density test. Rita's cortisol levels are elevated, but her thyroid hormone levels are normal. Her bone density is below normal, her fasting cholesterol is much too high, and the results of her glucose tolerance test indicate that she is diabetic.

3. What is Rita's preliminary diagnosis? Is there more than one possibility?

_____

_____

_____

Rita's physician orders a dexamethasone suppression test and an MRI of her head. Her results are within normal limits.

4. What is Rita's problem?

_____

_____

_____

5. What is the treatment?

_____

_____

_____

LEARNING ACTIVITIES

1. For each hormone list the effects on major systems. How many effects can you remember?

2. Using the Internet, investigate the effects of anabolic steroids. Can you explain why the side effects happen, given what you know about control of hormone levels?

3. Pick one steroid hormone and do some research. What are the effects? How wide ranging are they?

4. Play "Name That Hormone." One student should list information about a hormone while other students try to identify the hormone. How many clues do you need?

5. Play endocrine system "20 Questions." One student should pick a disease or an organ or a hormone. Other students ask questions requiring only "yes" or "no" answers to try to identify the term.

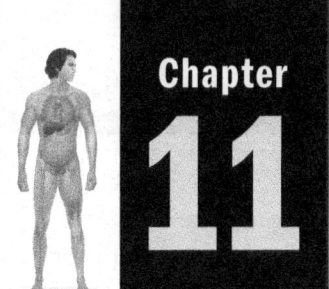

Chapter 11

THE SPECIAL SENSES: THE SIGHTS AND SOUNDS

CHAPTER SUMMARY

As we learned in the nervous system chapter, senses monitor the environment, collecting information and sending it into the nervous system. Senses include the special senses: vision, hearing, balance, taste, and smell, measured by specialized organs in the head, and general senses: touch, temperature, pain, hunger, and thirst, to name just a few, that are scattered around the body.

The organs of vision are the eyes, ball-shaped organs housed in the orbital cavities. The eye is protected by the eyelids, by a protective covering called the conjunctiva, and by tears secreted by the lacrimal glands. The eyeball has three layers. The outermost layer, the sclera, is the white part of the eyeball. Also contained in this layer is the cornea, a window that allows light into the eye. The middle layer, the choroid layer, is a pigmented layer. It also contains the iris, which controls the amount of light that enters the eye through the pupil. The innermost layer, the retina, is the layer of sensory cells (photoreceptors) that collect visual information. The interior of the eye is filled with fluid, which helps the eyeball keep its shape. Also in this layer is the lens, which focuses light on the retina. Infections, degenerative diseases, genetic disorders, and anatomical abnormalities can cause visual problems or even blindness. Some visual disorders can be corrected by lenses or surgery. Others have no reliable treatments.

The ear is the organ of hearing and balance. It can be divided into the external ear, the middle ear, and the inner ear. The external ear consists of the pinna, which collects sounds, and the auditory canal, which allows sound to pass into the ear. The eardrum (tympanic membrane) is found at the end of the canal. The middle ear starts at the eardrum and contains the auditory ossicles, the three smallest bones in the body. The bones transmit sound waves from the eardrum to the oval window. The Eustachian tubes, leading to the pharynx, are connected to the middle ear. The inner ear contains the semicircular canals (balance) and the cochlea (hearing). The cochlea is a fluid-filled chamber with hairlike receptor cells. When the fluid vibrates, the cells send the signal along the vestibulocochlear nerve to the brain. Hairlike cells in the semicircular canals react to changes in balance. They also transmit their signals along the vestibulocochlear nerve. Infections, conduction problems, and nerve damage can cause hearing problems.

The sense of taste is housed in the taste buds on the tongue. There are five types of taste: sweet, salt, bitter, sour, and umami. Taste is enhanced by the sense of smell, or olfaction. Olfactory receptors are found on the roof of the nasal cavity, behind the nose. Both taste and smell are chemical senses.

The sense of touch encompasses several different sensations, including vibration, temperature, pain, texture, and proprioception. Receptors for touch sensation are found mainly in the skin. More details on touch sensation were explained in Chapter 9.

CHAPTER OUTLINE

I. Types of senses

II. Vision
   A. External structures of eye
   B. Internal structures of eye
   C. Pathology: injuries and disorders of the eye
      1. Infections
      2. Visual disorders
      3. Degenerative disorders

III. Hearing
   A. Structure of the ear
      1. Outer ear
      2. Middle ear
      3. Inner ear and sound conduction
   B. Pathology: diseases of the ear
      1. Infection
      2. Progressive/chronic

IV. Other senses
   A. Taste
   B. Smell
   C. Touch
      1. Tactile
      2. Temperature
      3. Pain
      4. Proprioception

V. List of sensory disorders

VI. Pharmacology corner

MEDICAL TERMINOLOGY REVIEW

Define the following terms.

1. Cataracts: _____

2. Presbyopia: _____

3. Myopia: _____

4. Macular degeneration: _____

5. Glaucoma: _____

6. Vertigo: _____

7. Otitis media: _____

8. Labyrinthitis: _____

9. Phantom pain: _____

10. Referred pain: _____

 ## MULTIPLE CHOICE

Circle the letter of the correct answer.

1. Which of the following correctly describes *phantom* pain?
   a. Pain at the location where a vital organ was recently removed
   b. Pain that originates in one part of the body but is felt in another
   c. Pain that is mysteriously felt in the daytime but is elusive at nighttime
   d. Pain felt in a limb that was amputated

2. Umami is defined as:
   a. ringing in the ears.
   b. shadows in the visual spectrum.
   c. sporadic deafness and loss of equilibrium.
   d. taste of glutamates.

3. Which of the following is *not* a special sense?
   a. Vision
   b. Hearing
   c. Hunger
   d. Smell

4. How does the body rid itself of excess tears that are normally produced?
   a. The eyeball itself has the capacity to reabsorb the tears back into the aqueous humor.
   b. With each blink, the eyelid collects the tears and directs them to the back of the orbit.
   c. The face vasoconstricts its blood vessels, which increases the temperature of the eyeball and surrounding structures, and in turn evaporates the tears.
   d. Excess tears drain into the nose via two small holes in the inner corner of the eye.

5. Which of the following is true about the rods and cones?
   a. There are far more rods than cones.
   b. There are far more cones that rods.
   c. There are equal amounts of rods to cones.
   d. The number of rods and cones is correctable with prescription eyeglasses.

6. What is the primary function of the ossicles?
   a. Amplification of the sound waves that enter the middle ear
   b. Channeling of the sound waves that enter the outer ear
   c. Interpretation of sound waves that enter the inner ear
   d. Vibrates the eardrum

7. Which of the cranial nerves transmit from the cochlea and semicircular canals to the brain?
   a. Cranial nerve VIII
   b. Cranial nerve VI
   c. Abducens
   d. Both a and c

8. The eyeball sits in a conical cavity called the:
   a. optic.
   b. orbit.
   c. ocular.
   d. olfactory.

9. Which of the following structures function as sensors that activate a shielding effect as foreign objects approach the eyeball?
   a. Pupils
   b. Eyelashes
   c. Lens
   d. Eyebrows

10. Senses such as thirst, nausea, and the need to defecate are what kind of senses?
    a. Special
    b. Cutaneous
    c. Systemic
    d. Visceral

11. Which of the following is not a taste that the tongue's special sense can detect?
    a. Sweet
    b. Spicy
    c. Bitter
    d. Sour

12. Where is the eardrum located?
    a. Between the middle and the inner ear
    b. Between the middle and the outer ear
    c. Between the outer ear and the labyrinth
    d. At the outer rim of the external auditory meatus

13. The sense of taste is referred to as:
    a. olfactory.
    b. gastration.
    c. mechanoreception.
    d. gustatory.

14. Arrange the ossicles in the direction that sound waves would travel through them in order to be sent to the brain for interpretation:
    a. Malleus, incus, stapes
    b. Hammer, anvil, incus
    c. Stirrup, anvil, hammer
    d. Incus, malleus, stirrup

15. Which of the three layers of the eyeball is highly vascularized and also contains the iris?
    a. Cornea
    b. Choroid
    c. Retina
    d. Sclera

16. When there is low light, the iris will:
    a. defer activity to the rods.
    b. tighten.
    c. relax.
    d. rely on the cones.

17. The retina continues posteriorly to the back of the eye socket and forms what nerve?
    a. Oculomotor
    b. Optic
    c. Cranial nerve VIII
    d. Both b and c

18. The function of earwax is to:
    a. filter sound.
    b. trap foreign particles.
    c. maintain surface tension of the inner ear.
    d. monitor pressure.

19. What is the function of the muscles surrounding the lens of the eye?
    a. To alter the shape of the lens, making it either thinner or thicker
    b. To move the eyeball right, left, up, or down, depending on the focal point
    c. To decrease or increase the diameter of the iris
    d. To push the lens forward or pull it back, depending on the pressure of the fluids of the eye

20. When the iris contracts, what part of the eye changes, and in what way?
    a. Retina becomes wider
    b. Cornea becomes opaque
    c. Pupil becomes larger
    d. Pupil becomes smaller

21. Which one of the following taste receptors responds to glutamates?
    a. Sour
    b. Umami
    c. Salty
    d. Sweet

22. Where are the olfactory receptors?
    a. Back of throat
    b. Top of nasal cavity
    c. Sides of tongue
    d. In the organ of Corti

23. When Anne was 9 years old, she started having difficulty seeing the board from the back of the classroom. The teacher did not change the size of her writing to cause this gradual change in visual acuity. It was evident that Anne was in the early stages of:
    a. myopia.
    b. glaucoma.
    c. otis media.
    d. hyperopia.

24. Receptors of the skin, which include touch, heat, and pain, are referred to as:
    a. organ of Corti.
    b. visceral senses.
    c. cutaneous senses.
    d. dermatitis.

25. Sound travels best in:
    a. air at high altitudes.
    b. air at lower altitude.
    c. air at high temperature.
    d. solid or liquid medium.

26. An abscess that forms at the base of an eyelash is called:
    a. a stye.
    b. conjunctivitis.
    c. corneitis.
    d. a cataract.

27. Increased pressure in the fluid in the eye is this condition:
    a. cataract.
    b. glaucoma.
    c. conjunctivitis.
    d. presbyopia.

28. The medical term for nearsightedness is:
    a. hyperopia.
    b. presbyopia.
    c. myopia.
    d. amblyopia.

29. Macular degeneration results in the loss of _____ vision.
    a. peripheral
    b. color
    c. distance
    d. central

30. Poor night vision, _____, may be due to vitamin A deficiency.
    a. presbyopia
    b. red-green color blindness
    c. macular degeration
    d. None of the above

31. External otitis is more commonly known as:
    a. inner ear infection.
    b. vertigo.
    c. swimmer's ear.
    d. deafness.

32. _____ hearing loss occurs when sound waves are prevented from reaching the inner ear.
    a. Sensorineural
    b. Conductive
    c. Obstructive
    d. Temporal

33. A chronic, progressive disorder in which extra bone grows in the middle ear is called:
    a. otitis media.
    b. otosclerosis.
    c. external otitis.
    d. otonocia.

34. Inflammation of the inner ear, often caused by high fever, is called:
    a. otitis media.
    b. labrynthitis.
    c. Ménière's disease.
    d. otosclerosis.

35. Surgery to insert tubes through the tympanic membrane is known as:
    a. myringotomy.
    b. stapedectomy.
    c. labrynthotomy.
    d. tympanectomy.

MATCHING EXERCISES

Set 1

Please match each term with the appropriate definition.

_____ 1. Tympanic cavity
_____ 2. Hammer
_____ 3. Stirrup
_____ 4. Pinna
_____ 5. Anvil
_____ 6. Cochlea
_____ 7. Semicircular canals
_____ 8. Endolymph
_____ 9. Auditory tube
_____ 10. Vestibulocochlear

a. Canal or tube leading from the middle ear to the throat
b. Middle ossicle between the malleus and the stapes
c. Nerve carrying hearing and balance information
d. Outer ear
e. Fluid associated with organ of Corti
f. Ossicle directly against the oval window
g. Ossicle directly against the eardrum
h. Cranial nerve VI
i. Bony spiral structure of the inner ear associated with sound
j. Another name for the middle ear
k. Three canal loops of the inner ear associated more with equilibrium than actual sound

Set 2

Please match each disorder with the appropriate description.

_____ 1. Myopia
_____ 2. Labyrinthitis
_____ 3. Tinnitus
_____ 4. Conjunctivitis
_____ 5. Otitis media
_____ 6. Amblyopia
_____ 7. Presbyopia
_____ 8. Cataracts
_____ 9. Glaucoma
_____ 10. Hyperopia

a. Lazy eye
b. Inflammation of the covering of the eye
c. A ringing sound in the ears
d. Inflammation of the inner ear
e. Nearsightedness
f. Loss of taste
g. Farsightedness
h. Farsightedness brought about by age
i. Infection of the middle ear
j. Increased pressure in the fluid of the eye
k. Clouded lens of the eye

Set 3

Please match each structure with the appropriate definition.

_____ 1. Rods
_____ 2. Cornea
_____ 3. Cones
_____ 4. Sclera
_____ 5. Lens
_____ 6. Pupil
_____ 7. Lacrimal
_____ 8. Vitreous
_____ 9. Iris
_____ 10. Aqueous

a. Gland that secretes tears
b. Humor that bathes the iris, pupil, and lens
c. Bends light; surrounded by involuntary muscles
d. Clinical term for the entire middle layer of the eyeball
e. Sphincter that controls how much light passes into the eye
f. Humor that occupies the posterior cavity of the eyeball
g. Photoreceptors active in dim light
h. Hole or circular opening in the middle of the sphincter muscle of the eyes
i. Whites of the eyes
j. Photoreceptors active in bright light
k. Transparent structure allowing outside light rays into the eye

Set 4

Please match each disorder with the appropriate treatment.

_____ 1. Amblyopia
_____ 2. Presbyopia
_____ 3. Cataracts
_____ 4. Dry eye syndrome
_____ 5. Diabetic retinopathy
_____ 6. Red-green color blindness
_____ 7. Labyrinthitis
_____ 8. Tinnitus
_____ 9. Otosclerosis
_____ 10. Otitis media

a. Bifocals
b. Laser photocoagulation
c. Antibiotics
d. Eye patch
e. Avoid loud noises
f. Stapedectomy
g. Surgical correction of lens
h. Antivertigo medications, antihistamines
i. Avoidance, artificial tears
j. No treatment

FILL IN THE BLANK

Fill in the blanks to complete the following statements.

1. As it is associated with sound, the _____ of the inner ear sends sensory impulse to the _____ of the brain.

2. The inner ear is also called the _____.

3. Most of what you experience as taste is actually this sense:
_____.

4. The glands that produce tears are the
_____ glands.

5. Brown, hazel, blue, and green eyes are actually colors of the
_____ of the eyeball.

6. The fluid that fills the posterior cavity of the eye is called
_____.

7. Earwax is produced by the
_____ gland for the main

   purpose of
_____.

8. The vestibule chamber of the ear houses the
_____.

9. The part of the external ear that collects and directs sound waves into the external auditory meatus is the
_____.

10. The eardrum is clinically called the
_____ membrane.

11. The two fluids of the inner ear are the
_____ and the
_____.

12. The Eustachian tube leads from the ear to the
_____ of the throat.

13. As it is associated with equilibrium, the
_____ of the inner ear

    sends sensory signals to the
_____ of the brain.

14. An extremely contagious form of conjunctivitis is
_____.

15. Vision, hearing, and smell are known as
_____ senses.

16. The medical term for pinkeye is
_____.

17. Exposure to sunlight hastens the development of this eye disorder:
_____.

18. Diabetes mellitus may lead to
_____, a leading cause of

    blindness.

19. _____ is farsightedness
    that is caused by age.

20. This genetic visual disorder _____
    is much more common in males than females.

21. _____ is the medical term
    for dizziness.

22. Tinnitus is the feeling of _____
    in the ears.

23. Ménière's disease affects the

    _____.

24. Sensorineural hearing loss may result from damage to

    _____.

25. Organ pain felt on the body surface is known as

    _____ pain.

SHORT ANSWER

1. What are the functions of the eyelid?

    _____

    _____

    _____

    _____

2. Explain the phenomenon called *adaptation* as it applies to sensation.

    _____

    _____

    _____

    _____

3. Describe the process of *accommodation* as it applies to the lens of the eye.

    _____

    _____

    _____

    _____

4. What are the three layers of the eyeball?

    _____

    _____

    _____

    _____

5. Contrast the three types of auditory conduction.

    _____

    _____

    _____

    _____

6. Several visual disorders are more likely to develop as a person gets older.
    List and briefly describe those disorders.

    _____

    _____

    _____

    _____

7. Organ pain is often felt on the body surface. List the typical referred pain
    locations for your organs.

    _____

    _____

    _____

    _____

 LABELING ACTIVITIES

1. Label and color the parts of the eye using Figure 11–2 as your guide.

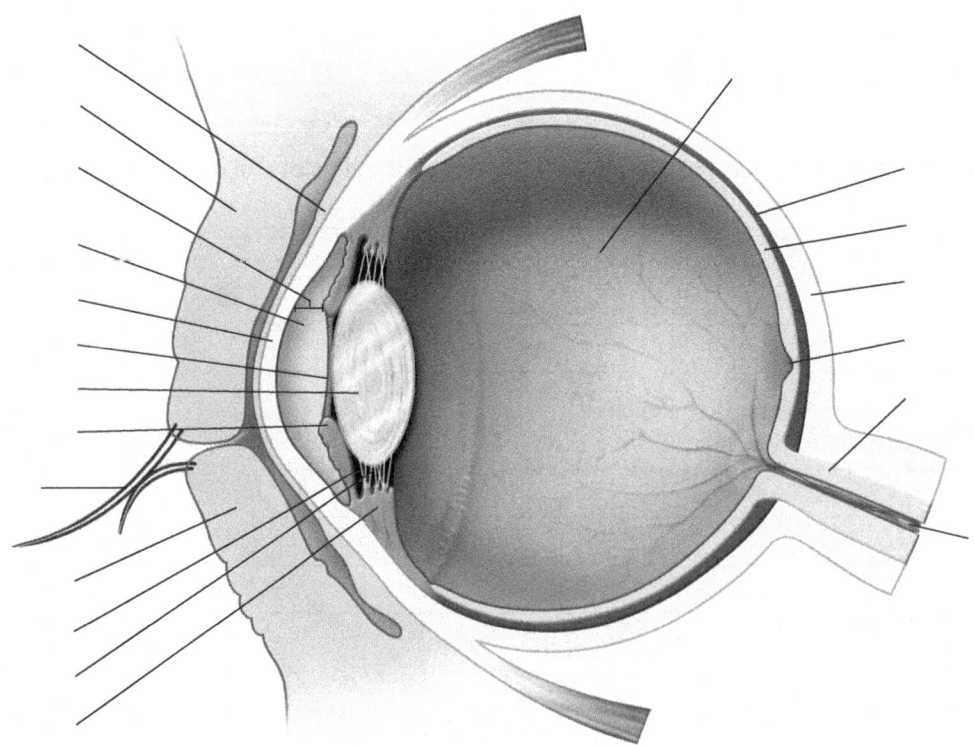

2. Label and color the parts of the ear using Figure 11–5 as your guide.

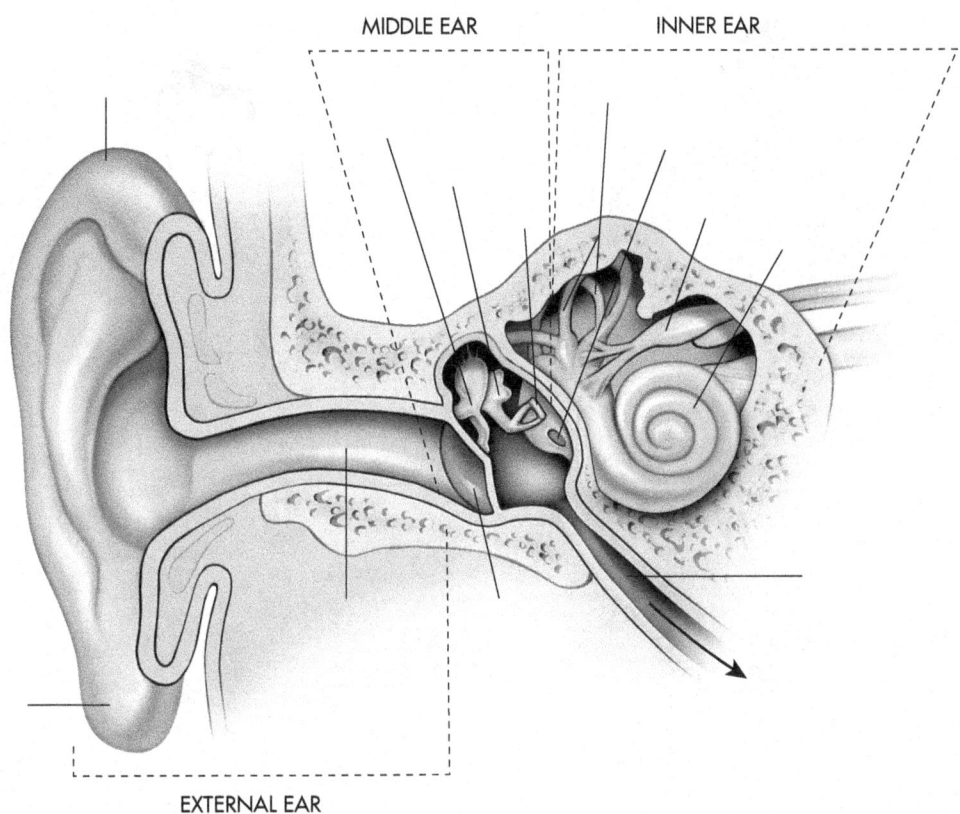

3. Label and color the referred pain map using Figure 11–10 as your guide.

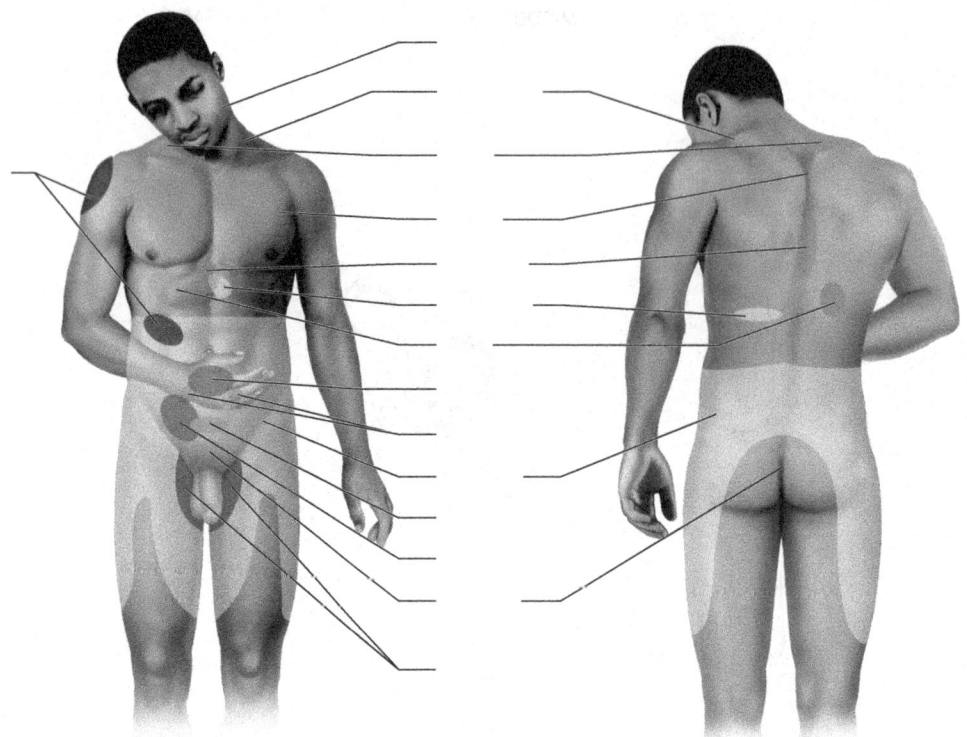

CASE STUDY

Ben turned 40 two years ago. For as long as he can remember, he has worn glasses, especially for driving and watching TV or movies. In the last couple of years his vision has deteriorated. His optometrist does a thorough eye exam and diagnoses Ben's problem.

1. Given Ben's age and the deterioration, what should Ben's optometrist rule out?

   _____

   _____

   _____

After all the tests, the optometrist tells Ben that his eyes are healthy, except for the recent changes in his vision. Ben needs bifocals.

2. What has caused the recent decline in Ben's vision?

   _____

   _____

   _____

LEARNING ACTIVITIES

1. Your sense of taste is dependent on your sense of smell. Using foods of similar texture (for example, potato and apple) test students' ability to identify items solely based on taste. Have students taste the food while holding their nose. Can they identify the food?

2. Your sense of touch is more sensitive on some parts of the body than on others. Explore this phenomenon using the two-point discrimination test. Have a student close his/her eyes. Gently touch two pins to their skin. Can the student tell that there are two pins? Move the pins closer together, then further apart. Test the student's ability to distinguish the pins on different parts of the body. Where can the pins be very close together? Where do they have to be further apart?

3. Using a Snellen eye chart, test your visual acuity. If you wear corrective lenses, try it without your lenses.

4. Put mystery objects into paper bags. Students should feel the object inside the bag without looking. Can they identify objects? Try objects that are similar to each other, like a peach in one bag and a nectarine in the other.

5. Using the Internet, research causes of blindness. What is the leading cause of blindness? What congenital or genetic disorders cause blindness?

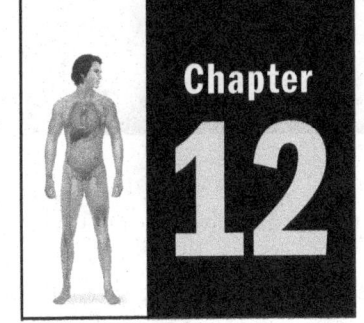
THE CARDIOVASCULAR SYSTEM: TRANSPORT AND SUPPLY

CHAPTER SUMMARY

The cardiovascular system consists of blood, the heart, and blood vessels. The heart pumps blood through the vessels around the entire body. The system is designed to move blood around the body, transporting much-needed oxygen, nutrients, and other chemicals to body tissues and carrying carbon dioxide and wastes away from the tissues.

The heart is a muscular pump located just to the left of center in the thoracic cavity. The heart is surrounded by a serous membrane, the pericardium. The heart wall has three layers, including a well-developed muscular layer. (Because the heart is a pump, it is constantly in motion.) The heart consists of four chambers: two upper chambers—the right and left atria and two lower chambers—the right and left ventricles. Between the chambers are valves: two atrioventricular valves (between the atria and ventricles) and two semilunar valves (between the ventricles and the vessels). Two walls, the interatrial septum and the interventricular septum, keep the blood on the right side of the heart separate from the blood on the left side of the heart.

Blood flows through the heart in a single direction. The superior vena cava and the inferior vena cava bring blood back from the body to the right atrium. The blood flows into the right ventricle (through the tricuspid valve) and then out the pulmonary trunk (thought the pulmonary semilunar valve) to the pulmonary arteries and on to the lungs to pick up oxygen and get rid of carbon dioxide. From the lungs, blood flows through the pulmonary veins back to the left atrium. From the left atrium, blood flows through the bicuspid valve and into the left ventricle. From there, blood flows through the aortic semilunar valve, out the aorta, and on to the body tissues.

The movement of the heart as it pumps blood around the body is known as the cardiac cycle. The cardiac cycle consists of four movements: atrial systole, atrial diastole, ventricular systole, and ventricular diastole. Systole, contraction, pumps blood out of a chamber, whereas diastole, relaxation, allows the chamber to fill with blood. The chambers contract and relax in coordination. The atria contract and pump blood into the ventricles, which are relaxed and filling, and then the ventricles contract and pump blood out of the heart. The atria relax during ventricular systole and begin to fill again. To keep blood flowing in the right direction, heart valves open and close at different times during the cardiac cycle. Atrioventricular valves are closed during ventricular systole and open during ventricular diastole. Semilunar valves are the opposite: closed during ventricular diastole and open during systole. The "lub dup" sound you hear when listening to a heartbeat is the sound of valves closing during the cardiac cycle, and the movement of blood due to the cardiac cycle can be felt at pulse points throughout the body.

Heart muscle cells are excitable cells like skeletal muscle and neurons. Thus, cardiac muscle contraction is associated with changes in electricity. But, unlike skeletal muscle, cardiac muscle does not have to wait for orders from the nervous system to contract. Although the nervous system can control heart activity, cardiac muscle contracts in its own rhythm, each cardiac muscle fiber essentially dancing to its own beat.

The cardiac conduction system ensures that all the muscle cells in the heart are synchronized so that the heart contracts and relaxes in a coordinated fashion so it can pump blood effectively. Pacemaker cells in the right atrium start the electricity flowing around the heart. From the sinoatrial (SA) node, electricity spreads to the right and left atria and to a second pacemaker, the atrioventricular (AV) node. From the AV node, electricity flows down the AV bundle to the left and right bundle branches (in the interventricular septum) and into the Purkinje fibers in each ventricle. In this manner electrical activity of the heart is coordinated. The electrical activity of the heart can be monitored using an EKG. Changes in the EKG suggest problems with the cardiac conduction system. Major changes are called arrhythmias.

Cardiac pathology can take many forms. Decreased blood flow to the heart muscle is called a heart attack or MI. The most common reason for MI is blockage of vessels supplying the cardiac muscle due to plaque formation. If blood flow is not quickly restored, the cardiac muscle will be irreversibly damaged. Damage to the cardiac muscle may cause failure of one or the other side of the pump, decreasing blood flow to the body and lungs or causing the activity of the heart to be uncoordinated. If heart valves are damaged, blood flow through the heart may be impaired. These kinds of valve problems are often diagnosed due to abnormal heart sounds called murmurs.

Blood has several functions, including transport, defense against infection, and body temperature regulation. Blood consists of two parts: plasma, a nonliving, liquid matrix, and the formed elements or blood cells, erythrocytes (red), leukocytes (white), and thrombocytes (platelets). Plasma is mainly a transport medium. Red blood cells carry oxygen, white blood cells defend against infection, and platelets cause blood clotting, preventing blood loss. Blood can be classified into four blood types, A, B, O, and AB, based on the antigens and antibodies present in the blood. Blood transfusions can only occur when blood types are compatible. AB blood is the universal receiver, and type O blood is the universal donor. Types A and B can never be mixed. If incompatible blood types are mixed, agglutination, clumping of red blood cells, will occur. That would be a potentially life-threatening condition. The presence (Rh positive) or absence (Rh negative) of Rh factor is also important. Blood disorders occur when the number of formed elements differs from normal. Either excess or decreased numbers of formed elements can cause significant health problems and are often a sign of major health issues. Blood clotting may also be abnormal for a number of reasons.

Blood is pumped through blood vessels. Blood vessel walls consist of several layers, including a layer of smooth muscle, which allows the vessels to change size to control blood flow to particular areas of the body. Large vessels that carry blood away from the heart are called arteries. Large vessels that carry blood toward the heart are called veins. Their smaller counterparts are called arterioles and venules, respectively. Embedded in tissues are tiny vessels running between arterial and venous circulation, called capillaries.

The pressure of blood flowing in the arteries can be measured as a way to monitor circulation. The ratio between systolic pressure, the pressure of blood flow during ventricular systole, and diastolic pressure, the pressure in the arteries during ventricular diastole, is known clinically as blood pressure (BP).

If a vessel is damaged, blood will leave the body, possibly causing dangerously low blood pressure. Blood loss is prevented by coagulation, or clotting. Coagulation is a series of biochemical reactions started by platelets, which plug any holes in blood vessels to keep blood from flowing out the hole. A clot is formed when the clump of platelets becomes insoluble in water, keeping it from being dissolved by plasma. There are many problems associated with blood vessels. Anything that could impede flow, weaken vessel walls, or either increase or decrease clotting ability could cause major problems. For a complete list of cardiovascular pathologies, see the list at the end of the chapter.

CHAPTER OUTLINE

I.   Overview of the cardiovascular system

II. The heart
   A. Structure and function
   B. Cardiac cycle
   C. Pathology
     1. Pump problems
     2. Vessel problems
   D. Electrical activity
     1. Electrical problems

III. Hematology
   A. Blood function
   B. Blood composition
     1. Plasma
     2. Cells
       a. Red blood cells
       b. White blood cells
       c. Platelets
   C. Inflammation
   D. Blood types
   E. Pathology

IV. Blood vessels
   A. Structure and function
   B. Capillaries and veins
   C. Blood pressure
   D. Blood clotting
   E. Pathology

V.   List of circulatory disorders

VI. Pharmacology corner

MEDICAL TERMINOLOGY REVIEW

Define the following terms.

1. Arteriosclerosis: _____

2. Thrombus: _____

3. Embolism: _____

4. Leukemia: _____

5. Myocardial infarction: _____

6. Peripheral vascular disease: _____

7. Aneurysm: _____

8. Anemia: _____

9. Heart block: _____

10. Arrhythmia: _____

MULTIPLE CHOICE

Circle the letter of the correct answer.

1. The white blood cells that produce heparin are:
   a. lymphocytes.
   b. basophils.
   c. eosinophils.
   d. erythrocytes.

2. The function of hemoglobin is:
   a. clotting.
   b. gas transport.
   c. immunity.
   d. production of blood cells.

3. Which of the blood types is considered the *universal recipient?*
   a. A
   b. O
   c. B
   d. AB

4. Where in the chest is the heart located?
   a. Directly in the middle of the chest, with apex above the base
   b. Slightly right of center with the base resting on the diaphragm
   c. Half to the right and half to the left of chest midline, with the base directly resting on the diaphragm
   d. Slightly left of center with the base above the apex

5. Which of the following is true about a type B-positive person?
   a. It is safe to give blood to a type B-positive and O-negative.
   b. It is safe to give blood to a type B-positive and AB-positive.
   c. It is safe to give blood to a type B-positive and O-positive.
   d. It is safe to receive blood from a type B-positive and type AB-positive.

6. The right side of the heart is responsible for:
   a. collecting and distributing both oxygenated and deoxygenated blood to and from the right side of the body.
   b. collecting deoxygenated blood from all over the body and sending it to the lungs.
   c. sending oxygenated blood to the upper body and collecting deoxygenated blood from the lower body.
   d. sending deoxygenated blood to and collecting oxygenated blood from the lungs.

7. What type of self-antigens do people with blood type O have?
   a. A antigens
   b. A and B antigens
   c. None
   d. O antigens

8. The left side of the heart is responsible for:
   a. collecting and distributing deoxygenated blood from and to the left side of the body.
   b. sending oxygenated blood to the lower body and collecting deoxygenated blood from the upper body.
   c. collecting oxygenated blood from the lungs and sending it to the entire body.
   d. sending deoxygenated blood to the lungs and collecting similar blood from the head.

9. What does serotonin do?
   a. Vasoconstricts
   b. Clots blood
   c. Unclots blood
   d. Increases temperature of the plasma

10. What types of plasma antibodies do people with blood type A have?
    a. anti A antibodies
    b. anti B antibodies
    c. anti O antibodies
    d. anti A and anti B antibodies

11. Which of the blood types is considered a universal donor?
    a. A
    b. B
    c. O
    d. AB

12. What prevents blood from shooting into the left atrium upon ventricular contraction?
    a. The third chamber, called the atrioventricular chamber
    b. A valve called the tricuspid
    c. A valve called the mitral
    d. Decompression of the diaphragm

13. Which of the following is true in regard to the Rh factor?
    a. If an Rh-positive father and an Rh-negative mother have a child that inherits the father's blood type, it will be healthy, but the second child of the same couple will have complications if he or she is also Rh-positive.
    b. If an Rh-negative father and an Rh-positive mother have a child that inherited the father's blood type, then there will be complications with the growth and development of this child.
    c. If an Rh-positive father and an Rh-negative mother have a child that inherits the mother's blood type, it will be healthy, but their second child, if Rh-positive, will have complications.
    d. If an Rh-positive father and an Rh-positive mother have an obvious Rh-positive child, it will be healthy, but their second child, if Rh-negative, will have complications.

14. How does chewing an aspirin tablet help in a heart attack?
    a. It conducts electrical current that gives the heart muscles an instant jolt.
    b. It has the ability to vasoconstrict, temporarily raising blood pressure.
    c. It has anticoagulating ability to help blood flow easier.
    d. It increases RBCs, allowing more oxygen to the heart muscles.

15. In measuring blood pressure, inflate the cuff to what measure? Then open the valve slightly so the cuff slowly deflates as you listen to what?
    a. 30 mm Hg above the point where you lose the pulse sound; brachial artery
    b. 120 mm Hg; radial artery
    c. 120 mm Hg above the point where you lose the pulse sound; carotid artery
    d. The point where you lose the pulse sound; chest

16. Which of the following statements is true?
    a. The right ventricle sends blood to both the right and left lungs to pick up a fresh supply of oxygen.
    b. The left ventricle sends blood to both the right and left lungs to pick up a fresh supply of oxygen.
    c. The right ventricle sends blood to the right lung, and left ventricle sends blood to the left lung, to pick up a fresh supply of oxygen.
    d. The right and left atria direct blood to the right and left lung, respectively.

17. Which of the heart chamber walls is the thickest?
    a. Atrioventricular
    b. Left ventricle
    c. Right atrium
    d. Left atrium

18. Which of the following is true about resting heart rate?
    a. On average, males have a faster rate than females.
    b. On average, male and female rates are the same provided they are the same weight.
    c. On average, females have a faster rate than males.
    d. There is no evidence that resting heart rate is different in males than in females.

19. Correctly arrange the electrical wiring of the heart from where the impulse is initially generated to where it is carried to the contractile muscle cells.
    a. AV bundle, vagus nerve, sino-atrial node, atrioventricular node
    b. Sino-atrial node, vagus nerve, Purkinje cells, atrioventricular node, AV bundle
    c. Sino-atrial node, atrioventricular node, AV bundle, Purkinje fibers
    d. Vagus nerve, sino-atrial node, atrioventricular node, Purkinje fibers, AV bundle

20. On the ECG, which of the waves represents the depolarization of the atria?
    a. T
    b. QRS
    c. It is masked by another wave.
    d. P

21. On the ECG, which of the waves represents the repolarization of the ventricles?
    a. T
    b. It is masked by another wave.
    c. QRS
    d. P

22. Immediately after the depolarization of the ventricles, what happens?
    a. The atria contract.
    b. The atria relax.
    c. The ventricles contract.
    d. The ventricles relax.

23. How much blood does a human normally have?
    a. 1 to 3 liters
    b. 4 to 6 liters
    c. 7 to 9 liters
    d. 11 to 13 liters

24. One of the leukocytes secretes heparin. What does heparin do?
    a. Clots blood
    b. Carries oxygen
    c. Threads a biological net
    d. Keeps blood from clotting

25. Which of the formed elements has the capacity to release serotonin?
    a. Platelets
    b. Basophils
    c. Eosinophils
    d. Neutrophils

26. Pain or discomfort in the chest due to lack of oxygen to the heart is known as:
    a. myocardial infarction.
    b. angina pectoris.
    c. heart failure.
    d. cor pulmonale.

27. Pulmonary edema is often caused by:
    a. left pump failure.
    b. right pump failure.
    c. myocardial infarction.
    d. angina pectoris.

28. The most common cause of secondary polycythemia is:
    a. cancer.
    b. poor nutrition.
    c. living at high altitudes.
    d. aging.

29. A clot formed by the sticking of platelets to blood vessel walls is called a(n):
    a. embolus.
    b. heart attack.
    c. thrombus.
    d. scab.

30. Atherosclerosis of vessels in the legs is called:
    a. peripheral embolic disease.
    b. peripheral thrombus disorder.
    c. peripheral appendicular clotting.
    d. peripheral vascular disease.

31. Abnormal heart sound due to leaking or narrowed valve is:
    a. prolapse.
    b. murmur.
    c. stenosis.
    d. All of the above

32. This often silent condition may lead to heart attacks and strokes:
    a. cor pulmonale.
    b. hypertension.
    c. myocardial infarct.
    d. pulmonary edema.

33. Swollen and distended superficial leg veins are called:
    a. hemorrhoids.
    b. atherosclerosis.
    c. peripheral vascular disease.
    d. None of the above

34. _____ is a genetic bleeding disorder.
    a. Hemophilia
    b. Thrombocytopenia
    c. Leukemia
    d. Polycythemia

35. This disorder would be treated by replacing the damaged heart valve:
    a. myocardial infarct.
    b. cor pulmonale.
    c. heart block.
    d. stenosis.

MATCHING EXERCISES

Set 1

Please match each term with the appropriate definition.

_____ 1. Monocytes
_____ 2. Phagocytosis
_____ 3. Erythrocytes
_____ 4. Basophils
_____ 5. Leukocytes
_____ 6. Lymphocytes
_____ 7. Hemopoiesis
_____ 8. Neutrophils
_____ 9. Thrombocytes
_____ 10. Eosinophils

a. Produce antibodies
b. The process by which RBCs are created
c. Granulocytes that attempt to destroy bacteria by engulfing
d. WBCs involved in allergies and inflammation
e. WBCs functioning to combat parasites
f. The process by which a cell surrounds and ingests an invader
g. Collective term for white blood cells
h. Collective term for red blood cells
i. Higher than normal amounts in chronic infections
j. Platelets

Set 2

Please match each term with the appropriate definition.

_____	1.	Agglutination
_____	2.	Pallor
_____	3.	Tunic
_____	4.	Embolus
_____	5.	Inotropism
_____	6.	Node
_____	7.	Anastomoses
_____	8.	Vasoconstriction
_____	9.	Vasodilation
_____	10.	Iron

a. Arterial reaction resulting in increased pressure within the vessel
b. Deficiency may lead to anemia
c. Pale skin
d. Self-antigens on RBC cell surface bind to antibodies, clumping
e. Branching of arteries so there is ample blood supply to the entire heart
f. A layer of a blood vessel
g. Results in increased blood vessel diameter
h. Pacemaker
i. Traveling clot
j. Force of cardiac contractions
k. Mineral produced by the heart

Set 3

Please match each disorder with the appropriate description.

_____	1.	Hemophilia
_____	2.	Cerebral vascular accident
_____	3.	Cor pulmonale
_____	4.	Anemia
_____	5.	Aneurysm
_____	6.	Myocardial infarction
_____	7.	Atherosclerosis
_____	8.	Arteriosclerosis
_____	9.	Leukemia
_____	10.	Valvular insufficiency

a. The right ventricle is not pumping blood efficiently
b. Accumulation of plaque in vessels
c. The mitral valve may be too large
d. High amounts of immature WBCs
e. Hardening of the blood vessels
f. Uncontrollable bleeding
g. Low amount of viable RBCs
h. Weakening of arterial wall
i. Heart attack
j. Stroke

Set 4

Please match each disorder with the appropriate treatment.

_____	1.	Cardiac tamponade
_____	2.	Angina
_____	3.	Embolism
_____	4.	Hypertension
_____	5.	Bradycardia
_____	6.	Endocarditis
_____	7.	Arteriosclerosis
_____	8.	Hemophilia
_____	9.	Leukemia
_____	10.	Aneurysm

a. Chemotherapy, bone marrow transplant
b. Clotting factor
c. Lifestyle changes, medication
d. Remove fluid
e. Clot busting drugs, surgery, preventative medications
f. Antibiotics
g. Observation, surgical repair
h. Nitroglycerin, oxygen, open coronary arteries
i. Pacemaker
j. Lifestyle changes, angioplasty, medication

FILL IN THE BLANK

Fill in the blanks to complete the following statements.

1. Arteries move blood _____ the heart.

2. The wall that separates the two lower chambers of the heart is called the _____.

3. On the ECG, the wave that represents the depolarization of the ventricles is the _____.

4. Blood transports

   _____,

   _____,

   _____, and

   _____.

5. Upon separation by centrifugation, blood is seen to have two major components, _____ and

   _____.

6. In the clotting mechanism, prothrombin, produced by the _____, is converted to thrombin with the help of vitamin

   _____.

7. The wall that separates the two upper chambers of the heart is called the _____.

8. Blood from the right upper chamber drains through the _____ valve to the lower chamber on the same side.

9. When body temperature increases, the response of the cardiac rate and force is to _____.

10. The SA node is located in/on the _____ of the heart.

11. The electrolyte _____, when at high levels, can prolong heart contractions to the point that the heart can actually stop beating.

12. The biological "net" or "gauze" made of _____ is formed by posttrauma attempts to cover the wound and prevent blood cells from escaping.

13. When examining the plaque removed from a blood vessel, you will find the main component of this substance is

   _____.

14. Two instruments, the

   _____ and the

   _____, are used to determine a patient's or client's blood pressure.

15. The two large vessels that empty into the right atrium are the
_____ and the
_____.

16. The medical term for a heart attack is
_____.

17. Buildup of plaque in blood vessels leading to reduced blood flow is known as _____.

18. _____ (hardening of the arteries) leads to brittle, less flexible blood vessels.

19. Rapid, uncoordinated contractions of heart muscle are called
_____.

20. Failure of impulses to flow smoothly around the heart, leading to uncoordinated heart activity is called
_____.

21. _____ is a decrease in the oxygen-carrying capacity of the blood due to decreased red blood cell number or decreased hemoglobin.

22. A cancer that results in increased numbers of white blood cells is called
_____.

23. A floating clot is called a(n)
_____.

24. A weak spot in the wall of an artery that could rupture causing a massive hemorrhage is called a(n)
_____.

25. John wakes his wife in the middle of the night because he is not feeling well. He is nauseous and dizzy. His left shoulder aches, and he is having trouble catching his breath. What is happening to John?
_____

SHORT ANSWER

1. Contrast the terms *agglutination* and *coagulation*.
_____
_____
_____
_____

2. Explain the exception to the rule that arteries carry oxygenated blood and veins carry deoxygenated blood.
_____
_____
_____
_____

3. Besides water, name four substances that are carried in blood.

_____

_____

_____

_____

4. What effects do the sympathetic and the parasympathetic divisions of the
    nervous system have on the heart, respectively?

_____

_____

_____

_____

5. How do the walls of the heart receive a blood supply to stay healthy?

_____

_____

_____

_____

6. Explain how lifestyle may influence cardiovascular health.

_____

_____

_____

_____

7. There are several disorders of blood cell number. List and briefly
    describe them.

_____

_____

_____

_____

LABELING ACTIVITIES

1. Label the figure and color the blood vessels using red to indicate oxygen-
    ated blood, blue to indicate deoxygenated blood, and purple to indicate
    where the oxygen level changes, using Figure 12–1 in your textbook
    as a guide.

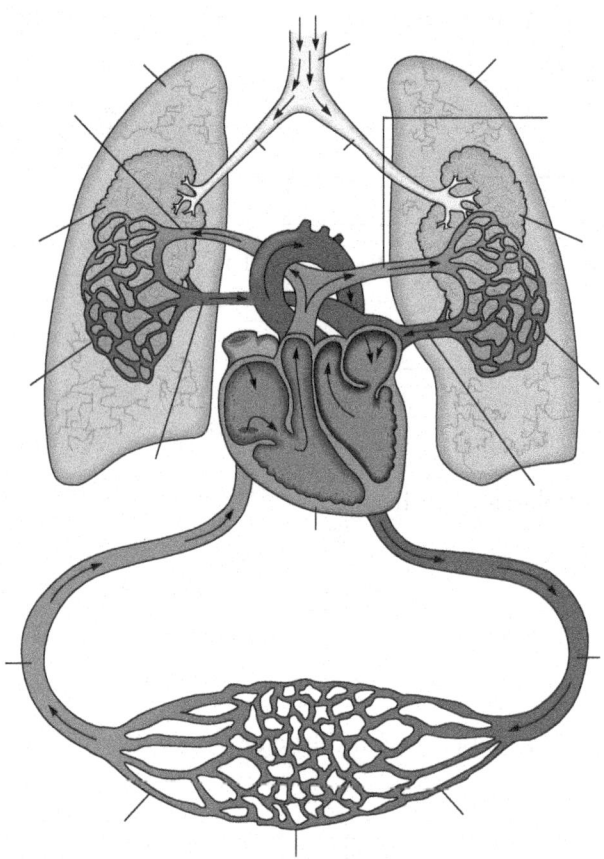

⬤ = Blood low in oxygen and high in
       carbon dioxide (deoxygenated).

⬤ = Blood high in oxygen and low in
       carbon dioxide (oxygenated).

2. Label the parts of the heart, using Figure 12–3 in your textbook as a guide. Insert the correct number in each circle.

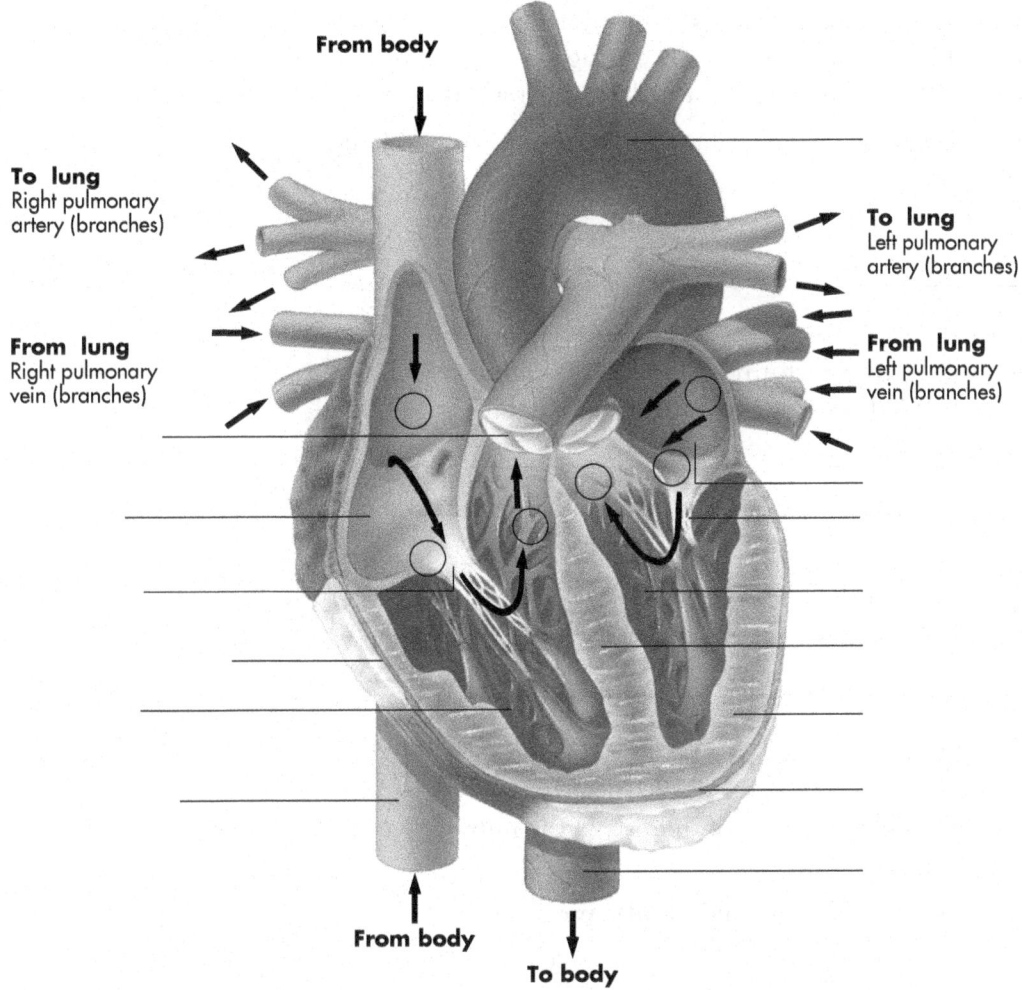

From body

To lung
Right pulmonary
artery (branches)

From lung
Right pulmonary
vein (branches)

To lung
Left pulmonary
artery (branches)

From lung
Left pulmonary
vein (branches)

From body

To body

RIGHT HEART PUMP

1. Deoxygenated blood returns from the upper and lower body to fill the right atrium of the heart creating a pressure against the atrioventricular (AV) or tricuspid valve.

2. This pressure of the returning blood forces the AV valve open and begins filling the ventricle. The final filling of the ventricle is achieved by the contracting of the right atrium.

3. The right ventricle contracts increasing the internal pressure. This pressure closes the tricuspid valve and forces open the pulmonary valve thus sending blood toward the lung via the pulmonary artery. This blood will become oxygenated as it travels through the capillary beds of the lung and then returns to the left side of the heart.

LEFT HEART PUMP

4. Oxygenated blood returns from the lung via the pulmonary vein and fills the left atrium creating a pressure against the bicuspid valve.

5. This pressure of returning blood forces the bicuspid valve open and begins filling the left ventricle. The final filling of the left ventricle is achieved by the contracting of the left atrium.

6. The left ventricle contracts increasing internal pressure. This pressure closes the bicuspid valve and forces open the aortic valve causing oxygenated blood to flow through the aorta to deliver oxygen throughout the body.

CASE STUDY

Jenny is a 35-year-old schoolteacher. She is active, participating in triathlons and going on ecotour vacations every year. Lately she has noticed that her stamina has decreased and that she is sometimes winded during an easy workout. She even had a dizzy spell. She decides to move up the date of her annual physical.

1. Many possible cardiovascular abnormalities could be the cause of Jenny's symptoms. List them.

    _____

    _____

    _____

2. During her physical, Jenny's doctor hears a gurgling sound when listening to Jenny's heart. What is the name for this disorder?

    _____

    _____

    _____

3. Jenny's history reveals that she had rheumatic fever as a child and was confined to bed for several days. What is probably Jenny's problem?

    _____

    _____

    _____

An echocardiogram confirms the diagnosis of stenosis of the mitral valve. Blood tests reveal that Jenny is also anemic.

4. What is the treatment for Jenny's cardiovascular problems?

    _____

    _____

    _____

5. What treatment might Jenny need in the future?

    _____

    _____

    _____

LEARNING ACTIVITIES

1. There are several types of leukemia, classified by the type of cells involved. Use the Internet to research the various types of leukemia. What cells are involved? Is the prognosis different for each type?

2. In 1986 a volleyball player named Flo Hyman died suddenly during a game. Her death was attributed to a ruptured aortic aneurysm due to Marfan's syndrome, a connective tissue disorder. Using the Internet or any other source, find out how Marfan's syndrome causes aneurysms.

3. Draw the cardiovascular system on a large piece of paper and divide it up into squares like a board game. Make a set of cards with questions on them. Every third space should have a symbol indicating that players must answer a question. If a player cannot answer the question, they must give up their next turn. The object of the game is to get from the right atrium through pulmonary and systemic circulation and back to the right ventricle. Make some cards worth extra places. Use dice to determine how many spaces should be moved each turn.

4. Using a model of the heart or a preserved heart, identify each of the parts. Using a pencil or probe, trace blood flow through the heart.

5. Play cardiovascular disorder "Jeopardy." On cards, write the name of a cardiovascular structure or disorder. Students must answer in the form of a question.

THE RESPIRATORY SYSTEM: IT'S A GAS

CHAPTER SUMMARY

To run cellular reactions, all our cells perform cellular respiration, a series of reactions that use up glucose to make ATP. To do cellular respiration our cells need lots of oxygen and give off lots of carbon dioxide as a waste product. Thus, there has to be a way to get lots of oxygen into the body and lots of carbon dioxide out. That is the job of the respiratory system. The respiratory system consists of a pair of lungs, upper and lower airways that conduct gas into and out of the system, air sacs called alveoli, a thoracic cage, and breathing muscles. It is important at the beginning of the journey through this system to point out the difference between respiration and ventilation. Respiration is actual exchange of gases. Internal respiration is gas exchange between tissues and blood, whereas external respiration is gas exchange between alveoli and blood. Ventilation, on the other had, is not gas exchange, but is the bulk movement of gases into and out of the lungs.

The respiratory system consists of airways (pipes) and lungs. The airways can be divided into upper and lower airways. The upper airways, also known as the upper respiratory system, function to heat or cool air to body temperature, filter particles, humidify air, provide the sense of smell, produce the voice, and conduct gas down to the lower airways (ventilation).

The nose is the entrance to the upper respiratory system. It has paired openings, the nostrils, and leads to a large cavity, the nasal cavity. The nasal cavity is divided into right and left halves by the nasal septum. The vestibular region is the entrance and contains nasal hairs that prevent particles from entering. The olfactory region, on the roof of the nasal cavity, contains the olfactory epithelium for your sense of smell. The major portion of the nasal cavity is the respiratory region. It contains a series of nasal conchae, bony shelves, which increase the surface area of the cavity. The functions of the nasal cavity include olfaction, humidifying air, warming air, and trapping particles. Several possible problems with the nasal cavity can interfere with ventilation, including nasal polyps and allergic rhinitis.

The upper airways are lined with a mucous membrane (respiratory mucosa) that acts as a mucociliary escalator. The mucociliary escalator consists of pseudostratified ciliated columnar epithelium coated with a thick layer of mucus. Particles get trapped in the mucus, and the cilia whisk them out of the airways. Connected to the nasal cavity are the sinuses. The frontal, maxillary, sphenoid, and ethmoid bones literally have holes in them. The holes, lined with respiratory mucosa, are called the paranasal sinuses. The sinuses are considered part of the respiratory system because their only opening is an opening into the nasal cavity. The sinuses lighten the head, provide resonance for your voice, and warm and moisten air.

After leaving the nasal cavity, air travels to the pharynx. The pharynx is divided into three parts: the nasopharynx, oropharynx, and laryngopharynx. The nasopharynx is the most superior part. It is meant to contain only inhaled air. In the nasopharynx are the adenoids (part of the lymphatic system) and the Eustachian tubes, which lead to the middle ear. The oropharynx is directly behind the mouth. Food and air travel through the oropharynx. In the oropharynx are the tonsils (part of the lymphatic system) and the soft palate and uvula, which block the nasal cavity during swallowing. The laryngopharynx is the most inferior part of the pharynx. It opens into both the respiratory and digestive systems.

From the pharynx, air passes into the larynx, the gateway to the lower respiratory system. The larynx, also known as the voice box, consists of a complex series of cartilages and ligaments. The thyroid cartilage, known as the Adam's apple, is the largest and most obvious cartilage. The epiglottis, at the superior end of the larynx, is a projection that folds over and covers the larynx during swallowing. In the interior of the larynx are the vocal cords. When the vocal cords touch and air is forced through them, voice is produced. The space between the cords is called the glottis. Air passes through the open glottis during inhalation. The vocal cords are the dividing line between the upper and lower airways.

Several infections can affect the upper airways. The common cold is a viral infection of the upper airways. Flu is also a viral infection of the upper airways, but it causes high fever and more systemic symptoms. Most parts of the upper airways can become inflamed from irritation or infection. The disorders include sinusitis, tonsillitis, pharyngitis, laryngobronchitis, and laryngitis. Bacterial infections may cause strep throat, or acute epiglottitis. Sleep apnea, interruption of breathing while asleep, is often caused by collapse of soft tissue in the back of the throat. As the tissues relax, they block the airway, blocking airflow. Loud snoring usually accompanies sleep apnea. Long-term sleep apnea can cause many health problems.

The lower respiratory tract begins at the vocal cords and can be thought of as a tracheobronchial tree because of the branching pattern of the pipes. The "tree" starts out as a single relatively large pipe, the trachea, and branches several times. With each branch, the pipes get smaller in diameter and more numerous. Initially the pipes are conducting pipes only, and no gas exchange takes place. Toward the end of the tree, gas exchange between lungs and blood takes place in the tubes. The tree starts with the trachea, which branches into two primary bronchi, one in each lung. Each primary bronchus branches into several secondary bronchi (lobar bronchi), one in each lobe. There are three lobar bronchi on the right and two on the left. Each lobar bronchus branches into several segmental bronchi. The bronchi then branch several more times until they form terminal bronchioles. (*Bronchiole* is a term meaning small bronchus.) The terminal bronchioles are the end of the conduction pipes. The terminal bronchioles branch into several respiratory bronchioles, which branch into alveolar ducts. Each alveolar duct has an alveolar sac and many alveoli at the end. The alveoli are the gas exchange sacs. In the respiratory bronchioles, alveolar ducts, sacs, and alveoli, gas exchange takes place.

Because the function of the tubes changes as you progress down the tracheobronchial tree, the anatomy of the tubes also changes. The trachea has walls made of C-shaped cartilage rings and is lined with respiratory mucosa. There are bands of smooth muscle at the posterior portion of each ring. Bronchi have cartilage plates, complete layers of smooth muscle, and are lined with respiratory mucosa. Bronchioles have no cartilage, reduced smooth muscle, and simple cuboidal epithelium. Finally, the alveoli have no cartilage, have no muscle, and are made solely of simple sqamous epithelium. If you remember that gas exchange must take place across the wall of the alveoli, the anatomical changes between the trachea and the alveoli make sense.

Gas exchange between alveoli and blood takes place across the alveolar capillary membrane. The membrane consists of a layer of fluid called surfactant, the simple squamous epithelium that makes up the alveolar walls (type I cells), the interstitial space (between the alveoli and capillaries), and the simple squamous epithelium that makes up the wall of the capillaries. Also in the alveoli are surfactant-secreting cells (type II cells) and macrophages (type III cells), which ingest foreign particles. Surfactant is secreted to prevent the delicate alveoli from being damaged by the movement of air into and out of them.

The health of your respiratory system can be tested using pulmonary function testing. Pulmonary function testing measures several different respiratory volumes. These volumes can then be used to calculate capacities. Each volume and capacity has a normal range of values. Deviation from the normal range indicates respiratory pathology of the lower airways. Pneumonia is an infection of the lungs that can be caused by many different pathogens. Tuberculosis is a bacterial infection that can remain dormant for years. It is highly contagious. Atelectasis is total or partial lung collapse. It can be caused by surgery, pain that prevents full ventilation, or injury. Perhaps the most important and common lung pathology is chronic obstructive pulmonary disease (COPD). It is a group of diseases in which patients have difficulty getting all the air out of their lungs. It is often accompanied by lung damage and increased secretions. The most common types of COPD are emphysema and chronic bronchitis. Asthma is an inflammatory condition often associated with allergies. During an asthma attack, some trigger causes the inflammation of airways, causing them to narrow. Triggers include cold air, exercise, allergies, and medication. Emphysema and chronic bronchitis are generally caused by smoking cigarettes. Emphysema is characterized by enlarged and/or destroyed alveoli. Chronic bronchitis is characterized by coughing, enlarged mucous glands, and increased smooth muscle in airways. Airways are inflamed, and lots of mucus is produced. Asthma is life-threatening but reversible with appropriate treatment. Patients with asthma usually take medication to prevent attacks.

Ventilation is accomplished because of the anatomy of the thoracic cavity. The thoracic cavity is surrounded by the thoracic cage, consisting of the ribs and sternum. The lungs are cone-shaped organs. The right lung has three lobes and the left only two because the heart is to the left of center in the thoracic cavity. The cavity itself contains several subcavities: the mediastinum, in the center of the thorax, the pericardial cavity, and two pleural cavities, one surrounding each lung. The pleural cavities are the potential space between the two layers of a serous membrane surrounding the lungs, the visceral pleura, covering each lung, and the parietal pleura, lining the thoracic cavity. Keep in mind, that like all serous membranes, the cavity is not really a cavity, but a potential space with a small amount of fluid. The two layers of the pleura are essentially glued together. Thus, the lungs are essentially glued to the inside of the thoracic cavity. There should be no space between the walls of the thorax and the lungs. If the pleural cavity is somehow breached or opened, problems result. A pneumothorax is air in the pleural or thoracic cavity. A pleural effusion occurs when there is excess fluid in the pleural cavity. If either air or fluid get into the thoracic cavity, pressure builds, which decreases the ability of the lungs to expand, eventually causing atelectasis.

Ventilation occurs due to pressure differences between the lungs and the air outside the body (atmosphere). Inspiration (inhalation) is an active process. Signals come from the medulla oblongata to the spinal cord and down paired phrenic nerves to the diaphragm, the chief muscle of inspiration. These signals stimulate the diaphragm to contract. When the diaphragm contracts, the volume of the thoracic cavity increases. Because the lungs are "glued" to the inside of the thoracic cavity, they too increase in volume. This increased volume decreases the pressure in the lungs, making it lower than atmospheric pressure. Air then flows into the lungs down the pressure gradient. Expiration, a passive process, occurs because the diaphragm relaxes. When the diaphragm relaxes, lung volume decreases, pressure in the lungs rises, and air flows out. It is a passive process because skeletal muscle relaxation is passive. Under certain conditions, expiration can become active when the abdominal muscles become involved, increasing the force of expiration. Expiration rate is controlled by carbon dioxide levels and blood pH. A list of respiratory disorders is found at the end of the chapter.

CHAPTER OUTLINE

I. Overview
   A. Basic anatomy
   B. Ventilation versus respiration
      1. Ventilation
      2. External respiration
      3. Internal respiration

II. The respiratory system
   A. The airways
      1. Upper airways
         a. General functions
         b. Nose
         c. Mucociliary escalator
         d. Sinuses
         e. Pharynx
         f. Larynx
         g. Pathology

2. Lower airways
    a. Tracheobronchial tree
    b. Alveolar capillary membrane
    c. Gas exchange
    d. Pulmonary volumes and capacities
    e. Pathology
        (1) Infections
        (2) Chronic obstructive pulmonary disease (COPD)
        (3) Asthma
B. The rest of the anatomy
    1. Pleura
    2. Lungs
    3. Thoracic anatomy
    4. Ventilation mechanism
C. List of diseases

MEDICAL TERMINOLOGY REVIEW

Define the following terms.

1. Atelectasis: _____

2. Emphysema: _____

3. Tuberculosis: _____

4. Pneumothorax: _____

5. Asthma: _____

6. Ventilation: _____

7. Respiration: _____

8. COPD: _____

9. Pleural effusion: _____

10. Compliance: _____

MULTIPLE CHOICE

Circle the letter of the correct answer.

1. What type of cells make up the membrane that lines the respiratory region of the nose and most of the airway?
    a. Stratified flagellated cuboidal
    b. Simple squamous
    c. Stratified globetulated squamous
    d. Pseudostratified ciliated columnar

2. The region that separates one lung from the other is:
    a. pleura.
    b. mediastinum.
    c. carini.
    d. septum.

3. The purpose of pleural fluid is to:
   a. reduce friction as an individual breathes.
   b. moisten air passage.
   c. filter debris.
   d. reduce surface tension within the bronchioles.

4. The main function of surfactant is to:
   a. nourish the trachea.
   b. reduce friction as a person swallows.
   c. reduce surface tension in the alveoli.
   d. filter gases in the nasal cavity.

5. Which of the following statements is/are true about sinuses?
   a. They connect to the nasal cavity via small passageways.
   b. We are born with three of the four sinuses.
   c. The sinuses are filled with air, making the skull heavier and more protective.
   d. All of the above

6. Which of the three sections of the pharynx contains the adenoids?
   a. Tracheopharynx
   b. Nasopharynx
   c. Oropharynx
   d. Laryngopharynx

7. Which of the following statements is true about inspiration?
   a. For inspiration to take place, pressure in the thoracic cavity needs to decrease.
   b. For inspiration to take place, atmospheric pressure needs to be lower than thoracic pressure.
   c. For inspiration to take place, pressure in the thoracic cavity needs to increase.
   d. For inspiration to take place, atmospheric pressure and thoracic pressure need to be equal.

8. What is the purpose of cilia in the airways?
   a. To propel air into the lungs
   b. To propel trapped debris upward to be expelled from the body
   c. To trap food and prevent entry into the windpipe
   d. For olfaction

9. Which of the three sections of the pharynx conducts air, food, and liquid?
   a. Oropharynx
   b. Nasopharynx
   c. Tracheopharynx
   d. All of the above

10. Which of the three parts of the sternum is fragile and needs to be avoided if performing CPR on a victim?
    a. Manubrium
    b. Body
    c. Hilus
    d. Xiphoid

11. Which of the paired tonsils is located in the middle pharyngeal section?
    a. Adenoid
    b. Palatine
    c. Lingual
    d. Submandibular

12. Directly below the Adam's apple is a large cartilage called the:
    a. carini.
    b. alveolus.
    c. philus.
    d. cricoid.

13. How does the epiglottis function in swallowing and breathing?
    a. As we breathe in, the epiglottis moves from its natural closed position to an open position so air can enter the larynx and trachea.
    b. As we swallow, the epiglottis flaps down to close off the larynx so food does not slip into that area.
    c. As we breathe in, the epiglottis closes off our esophagus so air does not enter into that area.
    d. It acts as a "guard gate," closing off the Eustachian tubes so air and food cannot enter and cause problems.

14. The name of the membrane that covers or wraps each lung is:
    a. pulmonary sheath.
    b. synovial membrane.
    c. visceral pleura.
    d. parietal aponeurosis.

15. The function of the conchae is to:
    a. warm and moisten air.
    b. filter large particles.
    c. trap oxygen so it remains in the airways.
    d. prevent the entrance of carbon dioxide into the airways.

16. Which of the regions of the nasal cavity houses the conchae?
    a. Respiratory
    b. Vestibular
    c. Olfactory
    d. Medulla

17. Inert gas, in the context of the respiratory system, means:
    a. it is poisonous to the continuation of life.
    b. it does not combine or interact in the body.
    c. it is necessary for sustaining life.
    d. it is depleting slowly from the atmosphere.

18. The process of gas exchange in which carbon dioxide is removed from the blood and oxygen added is called:
    a. internal respiration.
    b. internal ventilation.
    c. external respiration.
    d. external ventilation.

19. Arrange the following gases from highest to lowest percent in the atmosphere.
    a. Oxygen, carbon dioxide, nitrogen, argon
    b. Nitrogen, oxygen, carbon dioxide, argon
    c. Oxygen, carbon dioxide, argon, nitrogen
    d. Carbon dioxide, hydrogen, oxygen, argon

20. The largest source of oxygen released into the atmosphere is from the:
    a. ozone layer.
    b. fossil fuel.
    c. sahara desert.
    d. rain forest.

21. Where does the upper airway or upper respiratory tract end?
    a. Just below the nasopharynx
    b. Just below the vocal cords
    c. Just below the trachea
    d. Just behind the nasal cavity

22. What is the principal function of the vestibular region of the nasal cavity?
    a. Filter out large particles
    b. Internal respiration
    c. Smell
    d. Phonation

23. The bulk movement of air down to the lungs is termed:
    a. ventilation.
    b. respiration.
    c. transgasideous migration.
    d. pulmonary peristalsis.

24. Where is the olfactory region of the nose?
    a. Behind the nostril to the sides of the cartilage
    b. Against the septum of the nasal cavity
    c. The rear of the nasal cavity on the top of the uvula and soft palate
    d. Roof of the nasal cavity

25. Which of the following is not a function of the upper airway?
    a. Heating and cooling of inspired air
    b. Phonation
    c. Olfaction
    d. External respiration

26. Acute epiglottitis is caused by:
    a. smoking.
    b. bacterial infection.
    c. influenza virus.
    d. allergies.

27. Which of the following is not always a symptom of COPD?
    a. Sputum production
    b. Difficult expiration
    c. Decreased gas exchange
    d. Atelectasis

28. _____ is a condition in which the alveoli become enlarged.
    a. Tuberculosis
    b. Emphysema
    c. Asthma
    d. Cystic fibrosis

29. If blood pH drops (becomes more acidic), ventilation rate will:
    a. increase.
    b. stay the same.
    c. decrease.
    d. become uncontrolled.

30. Premature infants often have lung problems because they:
    a. are missing surfactant.
    b. have an immature nervous system.
    c. have not yet developed a diaphragm.
    d. have incomplete ribs.

31. An important public health issue that had faded from public attention due to antibiotic treatment and widespread testing, tuberculosis has become a potential problem again. Why?
    a. Foreign travel
    b. Antibiotic resistance
    c. Decreased vaccination in the United States
    d. Smoking

32. Respiratory distress, dysphagia, dysphonia, and drooling (the four Ds) are symptoms of:
    a. epiglottitis.
    b. tonsillitis.
    c. laryngitis.
    d. strep throat.

33. Which of the following is usually not an asthma trigger?
    a. Cold air
    b. Exercise
    c. Smoke
    d. Decreased blood pH

34. Patients with COPD should not receive 100% oxygen therapy. Why?
    a. They use oxygen levels to regulate ventilation.
    b. Oxygen is poison.
    c. Their phrenic nerves are damaged.
    d. All of the above

35. Asthma is often associated with this upper-respiratory disorder:
    a. laryngitis.
    b. epiglottitis.
    c. sinusitis.
    d. allergic rhinitis.

MATCHING EXERCISES

Set 1

Please match each technical term with the appropriate common term.

_____	1. Rectus abdominis	a.	Receptors of smell
_____	2. External intercostals	b.	Conchae
_____	3. Vibrissae	c.	Nose hair
_____	4. Vertebrocostal	d.	True ribs
_____	5. Phrenic	e.	Breastbone
_____	6. Medulla oblongata	f.	Nerve that innervates the diaphragm
_____	7. Turbinate	g.	Muscle of expiration
_____	8. Vertebrosternal	h.	Respiratory control center
_____	9. Sternum	i.	Muscle of inspiration
_____	10. Carina	j.	Where the trachea ends and primary bronchi begins
		k.	False ribs

Set 2

Please match each term with the appropriate description.

_____	1.	Capillary endothelium
_____	2.	Squamous pneumocytes
_____	3.	Terminal bronchiole
_____	4.	Alveolar epithelium
_____	5.	Interstitial
_____	6.	Pores of Kohn
_____	7.	Macrophages
_____	8.	Surfactant
_____	9.	Respiratory bronchioles
_____	10.	Granular pneumocytes

a. Coats the innermost layer of the alveoli
b. Marks the end of the conducting area of the lower respiratory tract
c. Leads to the alveolar ducts
d. Allows type III cells to move from one alveolus to another
e. Make up the majority of the actual tissue layer of the alveolus
f. Ingests foreign particles as they wander through the alveoli
g. Produce a phospholipid substance that acts on surface tension
h. Space that separates the alveoli from the capillaries
i. The actual tissue layer of the air sac functional units
j. Portion of alveolar capillary membrane that belongs to capillary

Set 3

Please match each term with the appropriate definition.

_____	1.	Pneumonia
_____	2.	Tuberculosis
_____	3.	Emphysema
_____	4.	Hydrothorax
_____	5.	Erythropoietin
_____	6.	Asthma
_____	7.	Pneumothorax
_____	8.	Atelectasis
_____	9.	Hemoglobin
_____	10.	Hemothorax

a. Molecule that carries large amounts of oxygen
b. Constriction of the airway in response to an allergy
c. Clinical term for the influenza virus
d. Infectious disease; vast lung damage can occur
e. Lung infection; inflammation with accumulation of cell debris and fluid
f. Blood in the pleural space
g. When the air sacs of the lungs are partially or totally collapsed
h. Air in the thoracic cavity
i. Hormone that influences RBC production
j. Irreversible condition in which air sacs become destroyed
k. Fluid accumulation in the pleural space

Set 4

Please match each disorder with the appropriate treatment.

_____	1. Allergic rhinitis	a. Radiation, chemotherapy, surgery
_____	2. Strep throat	b. Antibiotics
_____	3. Sleep apnea	c. Rest, fluids, treat symptoms
_____	4. Common cold	d. Mucus thinners, respiratory hygiene, digestive enzymes
_____	5. Asthma	
_____	6. Pneumothorax	e. Thoracentesis
_____	7. Emphysema	f. Weight loss, devices, surgery
_____	8. Cystic fibrosis	g. Antihistamines
_____	9. Chronic bronchitis	h. Bronchodilators, steroids, antihistamines
_____	10. Lung cancer	i. Oxygen, bronchodilators, enzyme replacement
		j. Antibiotics, bronchodilators, oxygen

FILL IN THE BLANK

Fill in the blanks to complete the following statements.

1. When air is breathed into the body via the nose, it is moistened to
   _____ relative humidity.

2. The three sections of the pharynx are the
   _____, the
   _____, and the
   _____.

3. The voice box is clinically known as the
   _____.

4. The common name for the trachea is the
   _____.

5. Approximately _____
   people die from smoking-related respiratory disease every year in the
   United States.

6. The region of the lung called the
   _____ is where pulmonary
   arteries exit, pulmonary veins enter, and the main stem of the bronchus
   can be found entering the lung.

7. The principal muscle of inspiration is called the
   _____.

8. The right lung has
   _____ lobes, and the left
   lung has _____ lobes.

9. A substance called surfactant can be found in the
   _____.

10. Both men and women have_____
    pairs of ribs, _____ of which
    are true ribs and _____
    are floating ribs.

11. Located just above the clavicle is the
    _____ of the lungs.

12. When thoracic volume increases, the thoracic pressure
    _____.

13. The gas exchange surface(s) of the lung is/are the
    _____.

14. When we exercise or participate in strenuous work, depth of breathing
    _____ and rate of
    breathing _____.

15. The Eustachian tubes lead from the
    _____ to the
    _____.

16. _____ are noncancerous
    growths within the nasal cavity.

17. Loud snoring may be a symptom of
    _____.

18. Inhaled foreign bodies more often end up in which lung?
    _____

19. The primary etiology for COPD is
    _____.

20. A hole in the lung will lead to
    _____.

21. The only method to slow progression of
    _____ is to stop smoking.

22. The most common cause of pneumothorax is
    _____.

23. A(n) _____ is a more
    permanent airway for ventilating patients with obstructed airways or who
    cannot breathe on their own.

24. Yelling at a rock concert can cause this disorder
    _____.

25. _____ is the fourth-leading
    cause of death in the United States.

SHORT ANSWER

1. Why is it a bad idea to get rid of nose hair?

    _____

    _____

    _____

    _____

2. What are the primary functions of the respiratory system?

    _____

    _____

    _____

    _____

3. Besides the diaphragm, what muscles play a role in inspiration? How do
    they aid inspiration?

    _____

    _____

    _____

    _____

4. Contrast internal and external respiration.

    _____

    _____

    _____

    _____

5. Explain why a foreign object is most likely to lodge in the right lung, if it
    gets past the larynx

    _____

    _____

    _____

    _____

6. Distinguish between asthma and COPD.

    _____

    _____

    _____

    _____

7. Explain why fluid or air in the pleural cavity may lead to atelectasis.

_____

_____

_____

_____

 LABELING ACTIVITIES

1. Label the various structures using Figure 13–1 in your textbook as a guide.

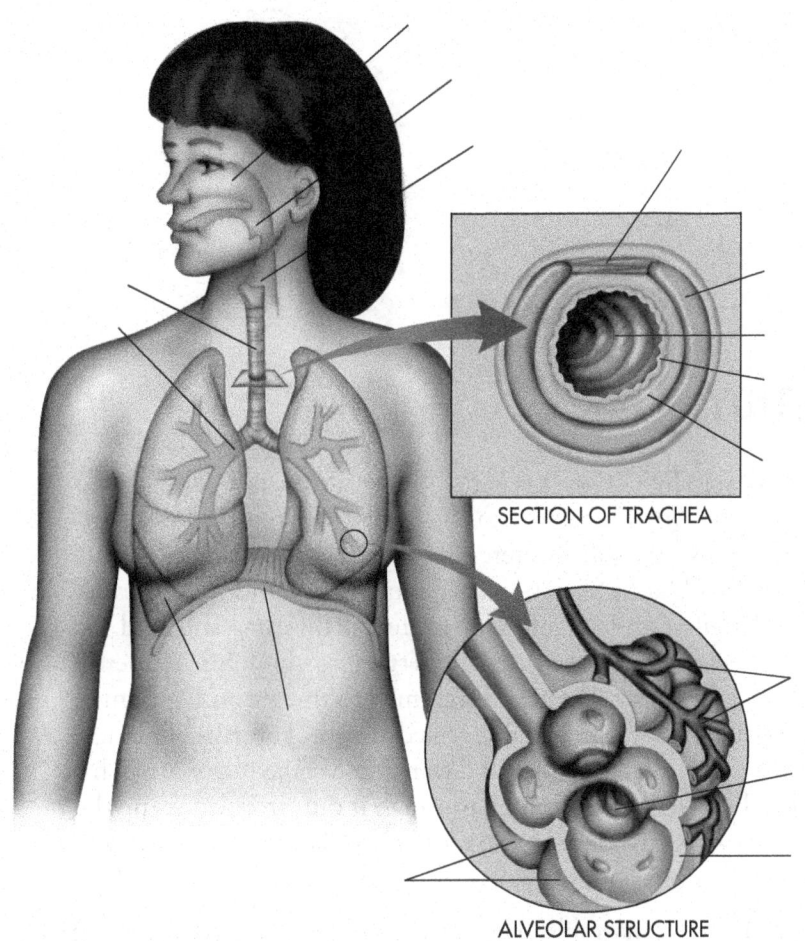

SECTION OF TRACHEA

ALVEOLAR STRUCTURE

2. Label the various structures using Figure 13–7 in your textbook as a guide.

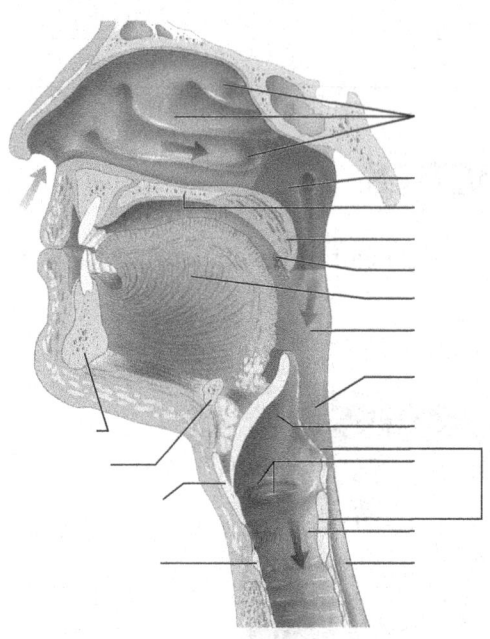

CASE STUDY

Ben is a 28-year-old, physically active, weekend sports enthusiast. He "rides" a desk all week at a downtown financial firm, but every weekend he rides his collection of "toys." He owns a personal watercraft, a Harley, a snowmobile, Rollerblades, a fancy mountain bike, a skateboard, a snowboard, and skis of every type. Physically fit and well versed in safety procedures, Ben has still had his share of injuries. Last weekend is a case in point. Trying to show his nephew a new skateboard trick, Ben lost his balance and slammed to the ground, groaning and gasping for air. X-rays at a local hospital revealed two broken ribs. The doctors sent Ben home with pain medication and a warning to take it easy until his ribs healed. As is typical for him, Ben decided to tough it out, stopped using the medication, and went to work two days after his injury. His only concession was riding the subway instead of his Harley.

Now, four days later, he feels miserable. He is in pain, coughing, and having trouble breathing. Struggling to breathe and coughing only increase the pain in his chest, which makes it harder to breathe. He is caught in a vicious cycle. He takes a dose of the pain medication that the doctors gave him and stays home from work, hoping that taking it easy for the day will give his ribs a break. He's sure he'll feel better tomorrow.

1. Should Ben be worried about his symptoms? Why?

    _____

    _____

    _____

Ben's buddy Joe hears that Ben is too sick to come to work. Alarmed, Joe goes to Ben's apartment during lunch hour. Ben is lying on the couch coughing and having obvious trouble breathing. Despite Ben's protests, Joe calls an ambulance. EMTs stabilize Ben and transport him to the hospital.

2. What are the possible diagnoses given Ben's recently broken ribs?

    _____

    _____

    _____

3. What tests should be taken to obtain an accurate diagnosis?

    _____

    _____

    _____

Tests reveal that Ben has hypoxia, no fever, rapid heart rate, and rapid breathing. Even with treatment he is distressed and in pain. A CT scan shows that Ben has developed a pneumothorax.

4. How should the injury be treated?

    _____

    _____

    _____

LEARNING ACTIVITIES

1. Use a balloon to demonstrate the relationship of volume to pressure changes. Fill the balloon with air. Squeeze the balloon to lower the volume. Eventually the balloon will pop. Did the pressure increase inside the balloon as you decreased the volume by squeezing, or did it decrease? What would happen if you could somehow make the balloon bigger without added air? Try the same thing with the bulb of a turkey baster. Squeeze the air out. As you stop squeezing and allow the bulb to expand (increasing the volume) what happens to pressure inside the bulb? Does air flow into the bulb or out?

2. Cigarette smoking causes a number of respiratory disorders. Use the Internet to research the possible long-term effects of cigarette smoking. How many disorders are linked to smoking?

3. Each part of the respiratory system has a unique function. For each part, list that function and one disorder that interferes with that function.

4. Make a set of flashcards with the name of a part of the respiratory system on one side and either the anatomy or function on the other. Quiz your partner.

5. Design a respiratory system board game. Draw the upper and lower airways and lungs. Each space should be associated with a question about a specific part of the respiratory system. Each player is an oxygen molecule. The object is to get from the atmosphere to the blood. Roll dice to determine how many spaces can be moved. Answering the question correctly allows movement on the board.

14

THE LYMPHATIC AND IMMUNE SYSTEMS: YOUR DEFENSE SYSTEMS

CHAPTER SUMMARY

The lymphatic and immune systems work together to defend your body against the invasion of pathogens. The lymphatic system is a second circulatory system, running parallel to the blood circulatory system. Fluid leaks from blood capillaries, flows through body tissues, and flows into lymphatic capillaries. As the fluid moves through the tissues, it washes any pathogens into the lymphatic capillaries. In addition, the system functions to recycle fluids lost from blood, to store and promote the maturation of some white blood cells, and to absorb lipids from the digestive system.

As mentioned earlier, the smallest pipes of the lymphatic system are lymphatic capillaries. They are similar in anatomy to blood capillaries and run parallel to them. The fluid in them is called lymphatic fluid. After passing through tissues, lymphatic fluid flows into lymphatic capillaries. From lymphatic capillaries, lymph flows into lymphatic vessels. Often lymphatic vessels will flow into lymph nodes. Lymph nodes are small bodies containing lymphatic tissue (lymphocytes and macrophages) surrounded by lymphatic fluid. As lymph flows into the lymph node, any pathogens contained in the fluid can be destroyed by the white blood cells. Fluid then flows out of the nodes via other lymphatic vessels and then into lymphatic trunks and one of two lymphatic ducts (right lymphatic duct and thoracic duct). From the lymphatic ducts, lymph flows into one of the subclavian veins. Lymph nodes are concentrated in several body regions, including the neck, armpits, anterior elbow, thoracic cavity, abdominal cavity, pelvic cavity, and groin. Tonsils and adenoids are patches of lymphatic tissue in the pharynx. There are also two lymphatic organs: the thymus, which is involved in maturation of lymphocytes, and the spleen, which functions to clean the blood of pathogens. Instead of being filled with lymph, the spleen is filled with blood.

Several disorders affect the lymphatic system, including tonsillitis, inflammation of tonsils caused by infection; mononucleosis, infection of the lymph nodes by a virus; and lymphoma, cancer of the lymphatic system. In addition, one of the major ways in which cancer spreads around the body is through the lymphatic system. Lymph node involvement is one criteria of the staging system used to evaluate some cancers.

The immune system is a complex series of mechanisms designed to prevent pathogens from entering your body or to destroy pathogens once they have entered. The whole system operates using a recognition system based on antigens (cell surface proteins) and antibodies (blood proteins that bind to antigens).

The immune system can be divided into two parts. The innate immune system is the first line of defense. It is inborn and cannot recognize or remember specific pathogens. Thus it does not improve with experience. The adaptive immune system improves with experience. It can remember and later recognize pathogens it has met before. Thus it can mount a more rapid response the second or third time it meets a pathogen.

There are many warriors in the fight against infection. Some are part of innate immunity, some are part of adaptive immunity, and some are part of both. Barriers prevent invaders from entering the body. There are physical barriers, like skin and mucous membranes, which are hard to penetrate and are packed with white blood cells and lymph capillaries. There are chemical barriers, fluids often associated with physical barriers, including tears, mucus, urine, saliva, and sweat. They often contain antibacterial chemicals. These barriers are found on body surfaces that come into contact with the environment on a regular basis. These barriers are part of innate immunity.

Cells are important weapons against invasion. These cells are leukocytes or modified leukocytes. Many cells are part of innate immunity. Neutrophils and macrophages (modified monocytes) gobble up pathogens and infected cells (phagocytosis). Esoinophils protect against parasites and basophils secrete chemicals that enhance immune response. Natural killer cells are lymphocytes that kill infected cells. Dendritic cells and macrophages also act as antigen-displaying cells. They are part of both innate and adaptive immunity. They wear the antigens of a pathogen on their surface and activate the B and T lymphocytes, the cells of adaptive immunity. Leukemia is a group of cancers in which leukocytes divide out of control, resulting in too many, mostly immature, white blood cells. Leukemia always increases the risk of infection.

In addition to chemical barriers, chemicals secreted by cells and tissues have an active role in immunity. Cytokines are proteins produced by damaged tissue and white blood cells. These increase inflammation, stimulate white blood cell production, and enhance phagocytosis. Some specific cytokines include interferon, the interleukins, and tumor necrosis factor. Complement is a specialized series of proteins in normal blood serum and plasma that, in combination with antibodies, cause the destruction of particulate antigens. The results of complement are rupture of bacteria, enhanced phagocytosis, clumping of cells, and stimulation of white blood cells. Complement and the cytokines are part of both innate and adaptive immunity.

Inflammation or inflammatory response occurs when tissue is damaged from infection or injury. In response to damage, tissues release chemicals that cause several changes. White blood cells are attracted to the site, blood capillaries become more permeable, releasing fluid into the area, and more blood flows to the site. The familiar symptoms of inflammation, redness, heat, swelling, and pain are caused by the effects of the chemicals released from damaged tissue. Inflammation is both good and bad. If there is too little inflammation, tissue heals very slowly. If there is too much inflammation, tissue may be further damaged by the inflammation itself.

Another weapon in the body's arsenal is fever. Fever is deliberate elevation of body temperature by the hypothalamus. It is part of innate immunity and is triggered by the release of cytokines and other chemicals. The hypothalamus resets the body's temperature set point higher than normal. The body responds by raising temperature to match the new set point, making the body less hospitable for pathogens. Fever, like inflammation, is both good and bad. If body temperature gets too high, tissue damage can result. If the temperature is not high enough, the pathogen will not be destroyed.

So, what happens when a pathogen enters your body? To enter, the pathogen must first get through physical and chemical barriers. If that happens, then neutrophils detect the presence of a foreign antigen. They cannot identify the antigen, just that it is "non-self." The neutrophils ingest the cell bearing the antigens, destroying the pathogen and/or the infected cell. The neutrophils also release chemicals to activate portions of the immune system. Macrophages and natural killer cells are stimulated by the chemicals released by neutrophils and soon arrive at the scene. Natural killers kill infected cells. Macrophages ingest infected cells and display their antigens. Both release chemicals to further enhance immune function, including trigger of complement cascade and inflammation. Pathogens are attacked by several different methods simultaneously.

To really destroy the pathogen, the adaptive immune system must get involved. Innate immunity is crude warfare. Healthy tissue is damaged in the process. Adaptive immunity spares healthy tissue, zeroing in on infected cells and pathogens precisely. Long before a pathogen invades the body, lymphocytes are sorted and selected. First, those cells that can bind to antigens and become active are selected to reproduce. This is positive selection. Then, of those lymphocytes that respond to antigens, those that attack self-antigens must be destroyed. This is negative selection. In order to work, adaptive immunity must target specific pathogens and ignore the body's own cells. If the immune system begins to attack self-antigens, an autoimmune disorder will result. Many body tissues can be attacked, including joints, the thyroid gland, the pancreas, and myelin. Autoimmune disorders are chronic, progressive, degenerative diseases. They are notoriously difficult to treat because treatment means turning off the immune system just the right amount.

When a pathogen is detected, lymphocytes must be activated and proliferate. Antigen-displaying cells, wearing the pathogen's antigen on the cell surface, patrol the lymph nodes searching for just the right T cell. When the antigen-displaying cell finds the right T cell, a helper T cell that matches the pathogen, the helper T cell will be activated by binding to the antigen-displaying cell. Both cells release chemicals, and the T cell "wakes up." This is the beginning of adaptive immunity. A few helper T cells are activated, but to fight thousands or millions of pathogens, just a few T cells are not enough. Activated helper T cells divide and reproduce until there are thousands of them, all matching the pathogen. This is lymphocyte proliferation. Once helper T cells have proliferated, it is their job to activate B cells and other types of T cells. Without helper T cells, adaptive immunity never becomes active. This is the main manifestation of acquired immune deficiency syndrome (AIDS), caused by the human immunodeficiency virus (HIV). HIV preferentially attacks helper T cells, eventually destroying them. Without helper T cells, other lymphocytes are never activated, and adaptive immunity disappears. Without adaptive immunity, the body is basically wide open to pathogens. Simple infections become deadly, and some infections that would normally never take hold get an easy entrance to the body. Patients with full-blown AIDS generally die of infection, though other complications are also likely.

The job of helper T cells is to activate other lymphocytes, B cells, and several types of T cells. Helper T cells activate other lymphocytes by binding to the matching lymphocyte and releasing chemicals to stimulate activation and proliferation. B lymphocytes are responsible for a type of adaptive immunity known as antibody-mediated immunity. B cells called plasma cells release antibodies into the bloodstream that bind to and inactivate specific antigens. The effects of antibodies include neutralizing the antigen, clumping of antigens, activation of complement, immune stimulation, and enhanced phagocytosis. Thus, B cells destroy the pathogen and infected cells specifically and further stimulate both innate and adaptive immunity. B cells also can differentiate into memory B cells, which hang around in the body waiting for the pathogen sometime in the future. They are responsible for the secondary immune response, which is much faster and bigger than the first meeting with the pathogen.

There are several different types of T cells. Cytotoxic T cells, which are directly activated by antigen-displaying cells, are responsible for cell-mediated immunity, killing infected cells by contact. Cytotoxic T cells release a chemical that literally blows a hole in the infected cell, destroying it. Like plasma cells, cytotoxic T cells also release immune-stimulating chemicals. Regulatory T cells are part of the control system of the immune system. How they regulate the immune system is not well understood, but they function somehow as an off switch. Memory T cells are also made when T cells are activated. Like memory B cells, memory T cells wait in preparation for another invasion.

Adaptive immunity is acquired immunity; it is affected by experience. Acquired immunity can be natural (without medical intervention) or artificial (with medical intervention) and active (the immune system actively fights) or passive (the immune system does little work). For example, natural active immunity occurs with exposure to a pathogen, whereas natural passive immunity occurs when antibodies are passed from mother to child in utero or in breast milk. Artificial active immunity occurs due to immunization; artificial passive immunity occurs when antibodies are injected into a patient to help them fight off infection their immune system is incapable of fighting.

There are two major things to keep in mind when looking at the big picture of the immune system: (1) it is all about positive feedback, and (2) innate and adaptive immunity mutually enhance each other. Once an immune response is triggered, both innate and adaptive will get involved, and once turned on, the whole thing is hard to turn off. For a list of disorders, both of overactive and underactive immunity, please see the list at the end of the chapter.

CHAPTER OUTLINE

I. Overview

II. Lymphatic system
   A. Functions
   B. General anatomy
      1. Lymphatic circulation
      2. Lymph nodes
   C. Lymph organs
   D. Pathology
      1. Tonsillitis
      2. Lymphatic disorders
      3. Cancer stages

III. Immune system
   A. Antigens and antibodies
   B. Innate versus adaptive immunity
   C. Components
      1. Barriers
      2. Cells
         a. Types
         b. Leukemia
      3. Chemicals
      4. Inflammation
      5. Fever

IV. How the immune system works
   A. Innate immunity
   B. Adaptive immunity
      1. Lymphocyte selection
         a. Autoimmunity
      2. Lymphocyte activation
      3. Lymphocyte proliferation
      4. HIV and AIDS
      5. Action
         a. B cells
            (1) Plasma cells—antibodies
            (2) Memory B cells
         b. T cells
            (1) Helper T cells
            (2) Cytotoxic T cells
            (3) Regulatory T cells
            (4) Memory T cells

V. The big picture

MEDICAL TERMINOLOGY REVIEW

Define the following terms.

1. Anaphylaxis: _____

2. Cytokine: _____

3. Antibody: _____

4. Antigen: _____

5. AIDS: _____

6. Autoimmune disorder: _____

7. Innate immunity: _____

8. Adaptive immunity: _____

9. Lymph node: _____

10. Leukemia: _____

MULTIPLE CHOICE

Circle the letter of the correct answer.

1. Which stage of cancer is often terminal?
    a. IV
    b. III
    c. II
    d. I

2. Barriers of the immune system include:
    a. skin.
    b. mucous membranes.
    c. saliva.
    d. All of the above

3. How do antibodies destroy pathogens?
    a. May cause the antigens to clump
    b. Pinocytosis
    c. Phagocytois
    d. Pull the antigen to the body surface, ulcerate the skin, and then release the antigen to the external environment

4. What do the lymphatic system, innate immunity, and adaptive immunity have in common?
    a. Rid the body of invading pathogens
    b. Lay dormant until needed
    c. Never turn on themselves
    d. Get stronger and better with age

5. After the physical barriers, which of the following is considered the first line of defense in the body?
    a. Antibodies
    b. Histamines
    c. Heparin
    d. Phagocytosis

6. The function of the spleen is to:
    a. produce red blood cells.
    b. differentiate T lymphocytes.
    c. help WBCs mature.
    d. filter pathogens from the bloodstream.

7. The right lymphatic duct empties into the:
    a. thoracic duct.
    b. jugular vein.
    c. subclavian vein.
    d. spleen.

8. How does lymph move through the body?
    a. Body movement
    b. Heart
    c. Gravity
    d. Centrifugal force

9. Lymphatic trunks empty into:
    a. collecting ducts.
    b. subclavian veins.
    c. lymph nodes.
    d. thymus.

10. Where is the spleen located?
    a. Between the heart and the sternum
    b. Upper right quadrant of abdomen
    c. Upper left quadrant of pelvis
    d. Upper left quadrant of abdomen

11. Which of the following areas have large concentrations of lymph nodes?
    a. Lumbar
    b. Subclavian
    c. Inguinal
    d. All of the above

12. Where do lymphocytes originate?
    a. Yellow bone marrow
    b. Red bone marrow
    c. Spleen
    d. Lymph nodes

13. Which one of the WBCs does the human immunodeficiency virus specifically target?
    a. Helper B cells
    b. Helper T cells
    c. Plasma cells
    d. Macrophages

14. Antibodies passed on to a fetus through the placenta represent:
    a. innate immunity.
    b. naturally acquired passive immunity.
    c. artificially acquired passive immunity.
    d. naturally acquired active immunity.

15. Lymph from the lower extremities will eventually empty into the:
    a. femoral vein.
    b. inferior vena cava.
    c. right axillary vein.
    d. None of the above

16. A vaccine is an example of:
    a. innate immunity.
    b. artificially acquired active immunity.
    c. artificially acquired passive immunity.
    d. naturally acquired active immunity.

17. Which of the following is/are lymphatic trunks?
    a. Axillary
    b. Jugular
    c. Cervical
    d. All of the above

18. Which of the following is true about the spleen?
    a. It is not a vital organ.
    b. It produces red blood cells.
    c. As we age, we rely on the spleen more.
    d. All of the above are true.

19. Which WBC is the first to arrive at the site of damage?
    a. Macrophage
    b. Plasma
    c. Neutrophil
    d. Lymphocyte

20. Biological increase in body temperature due to infection represents:
    a. innate immunity.
    b. naturally acquired passive immunity.
    c. artificially acquired passive immunity.
    d. naturally acquired active immunity.

21. Injection of antibiotics like penicillin is an example of:
    a. naturally acquired active immunity.
    b. naturally acquired passive immunity.
    c. artificially acquired passive immunity.
    d. None of the above

22. Which of the following vessels collects two-thirds of the body's lymph?
    a. Inguinal duct
    b. Thoracic duct
    c. Right lymphatic duct
    d. Splenic duct

23. Which of the following WBCs are the most common in the bloodstream?
    a. Lymphocytes
    b. Monocytes
    c. Basophils
    d. Neutrophils

24. Where is the thymus located?
    a. Neck
    b. Chest
    c. Upper abdomen
    d. Brain

25. Which of the following is true about the function of the thymus gland?
    a. It has a higher functioning capacity in children than adults.
    b. It contains lymphocytes.
    c. It secretes a hormone.
    d. All of the above

26. Infection with Epstein-Barr virus causes:
    a. Hodgkin's Disease.
    b. mononucleosis.
    c. AIDS.
    d. lymphadenitis.

27. In lupus erythematosis, the immune system attacks:
    a. joint linings.
    b. myelin.
    c. thyroid gland.
    d. tissue in general.

28. Cancer that has spread to the lymph nodes is in stage:
    a. I.
    b. II.
    c. III.
    d. IV.

29. Leukemia in which stem cells are dividing out of control and symptoms are rapid and severe is called:
    a. acute myelogenous leukemia.
    b. acute lymphocytic leukemia.
    c. chronic myelogenous leukemia.
    d. chronic lymphocytic leukemia.

30. An autoimmune disorder in which the immune system attacks the beta cells of the pancreas is:
    a. multiple sclerosis.
    b. rheumatoid arthritis.
    c. Graves' disease.
    d. diabetes mellitus (type 1).

31. Drugs that suppress immune-enhancing chemicals and are used for treating autoimmune disorders are called:
    a. disease modifying antirheumatic drugs.
    b. biological response modifiers.
    c. nonsteroidal anti-inflammatory drugs.
    d. therapeutic steroids.

32. An autoimmune disorder in which the joints are destroyed is:
    a. lupus.
    b. gout.
    c. osteoarthritis.
    d. None of the above

33. Predictable worsening of allergies in patients with repeated exposure to allergens is called:
    a. atopic march.
    b. asthma.
    c. anaphylaxis.
    d. All of the above

34. Severe reduction in T cell numbers due to a genetic disorder is called:
    a. AIDS.
    b. RA.
    c. SCID.
    d. CML.

35. Inflammation is often thought of as a two-edged sword. Why?
    a. Too much inflammation can suppress the immune system.
    b. Too much inflammation encourages infection.
    c. Too much inflammation causes leukemia.
    d. Too much inflammation causes tissue damage.

MATCHING EXERCISES

Set 1

Please match each term with the appropriate definition.

_____ 1. Leukocyte
_____ 2. Neutrophil
_____ 3. Interferon
_____ 4. Macrophage
_____ 5. Basophil
_____ 6. Eosinophil
_____ 7. Natural killer cell
_____ 8. Dendritic cell
_____ 9. Cytotoxic T cell
_____ 10. Plasma cell

a. Phagocytic granulocytes; most common WBC
b. Release chemicals to promote inflammation
c. All-encompassing term for white blood cells
d. Adaptive immunity T-lymphocyte
e. Cytokine that protects neighboring cells from viral attack
f. Phagocytic modified monocytes; innate immunity
g. Produces antibodies
h. Active during parasitic infections
i. Modified monocytes acting as antigen-displaying cells
j. Innate lymphocytes that secrete chemicals to kill cells displaying antigens

Set 2

Please match each function with the appropriate term.

_____ 1. Primary immune response
_____ 2. Secondary immune response
_____ 3. Cell-mediated immunity
_____ 4. Turn off immune response
_____ 5. Natural active immunity
_____ 6. Artificial passive immunity
_____ 7. Artificial active immunity
_____ 8. Natural passive immunity
_____ 9. Innate immunity
_____ 10. Treats rheumatoid arthritis

a. Memory B cells
b. Plasma cell
c. Perforin
d. Regulatory T cells
e. Tumor necrosis factor inhibitor
f. Physical barrier
g. The flu shot
h. Accidental exposure to a pathogen like chicken pox
i. Being injected with antibodies
j. Breast milk

Set 3

Please match each term with the appropriate description.

_____ 1. Memory cells
_____ 2. Histamines
_____ 3. Interleukin-1
_____ 4. Lymph
_____ 5. Antigens
_____ 6. Fungi
_____ 7. Radiation
_____ 8. Venoms
_____ 9. Interleukin-2
_____ 10. Antibodies

a. Secreted by helper T cells
b. Located on cell surface
c. Remembers pathogens
d. Secreted by macrophages
e. Pathogenic organism
f. Inflammation-causing physical agent
g. Inflammation-causing chemical agent
h. Secreted by mast cells
i. Carries antigens to nodes around the body
j. Secreted by the plasma cells

Set 4

Please match each disorder with the appropriate treatment.

_____	1. Lymphoma	a. Enzyme replacement, sterile environment, gene therapy, bone marrow, or stem cells
_____	2. Tonsillitis	
_____	3. Rheumatoid arthritis	b. Insulin injections, pancreas transplant
_____	4. HIV/AIDS	c. Drug cocktails, treat infections
_____	5. SCID	d. Chemotherapy, radiation, bone marrow transplant
_____	6. Allergies	
_____	7. Mononucleosis	e. NSAIDs, steroids
_____	8. Type 1 diabetes	f. Antihistamines, avoidance, steroids
_____	9. Anaphylaxis	g. DMARDS, BRMS, surgery, lifestyle changes
_____	10. Inflammation due to injury	h. Rest, fluids, pain relievers
		i. Antibiotics if bacterial infection
		j. Epinephrine

FILL IN THE BLANK

Fill in the blanks to complete the following statements.

1. When your body mounts a hyperactive response to a harmless antigen, the reaction is called a(n)

   _____.

2. The lymphatic tissue inside the lymph nodes contains

   _____ and

   _____.

3. The destruction of self-destroying lymphocytes is known as

   _____.

4. Helper T cells are also called

   _____ cells.

5. To fight off thousands of pathogens, lymphocytes must make thousand of copies of themselves. This process is called lymphocyte

   _____.

6. A hypersensitivity reaction called

   _____ leads to

   _____ blood pressure

   and heart failure.

7. The thoracic duct empties into the

   _____.

8. The _____ produces a hormone that stimulates lymphocyte production in children.

9. When cancer has spread to nearby lymph nodes, it is in stage

   _____.

10. The part of the brain that regulates body temperature is the

    _____.

11. The lymphatic tissue inside a lymph node is surrounded by
_____.

12. During an allergic reaction, pollen directly activates the release of
_____.

13. Redness, swelling, heat, and possible pain are all
_____ symptoms.

14. The formed elements in whole blood consist of erythrocytes, platelets,
and _____.

15. Interferon and interleukins are both chemicals collectively called
_____.

16. Infection of the tonsils is called
_____.

17. Cancer that has spread far from the site of origin is called stage
_____.

18. In _____
leukemia, lymphocytes reproduce out of control.

19. You ice an injury to decrease
_____.

20. Myasthenia gravis is caused by autoimmune attack on
_____.

21. _____ cells are attacked by
the HIV virus.

22. The _____ drugs are
highly toxic and confusing pharmacologically but are still often used to
treat autoimmune disorders.

23. David, known as the "boy in the plastic bubble," had to spend most of
his life in a sterile environment because of this genetic disorder
_____.

24. _____ drugs suppress the
immune system.

25. Cancer of the lymph nodes is called
_____.

SHORT ANSWER

1. Besides HIV that causes AIDS, how can a patient's immune system become compromised?

    _____

    _____

    _____

    _____

2. What is the primary function of lymph nodes?

    _____

    _____

    _____

3. Structurally contrast the spleen and large lymph nodes.

    _____

    _____

    _____

    _____

4. What is the purpose of a fever?

    _____

    _____

    _____

    _____

5. Why are CD 4 cells so important?

    _____

    _____

    _____

    _____

6. Compare and contrast DMARDs and BRMs.

    _____

    _____

    _____

    _____

7. Explain the symptoms of systemic lupus erythematosis.

    _____

    _____

    _____

    _____

LABELING ACTIVITIES

Label the various parts of the lymphatic system using Figures 14–1 and 14–2 in your textbook as your guide.

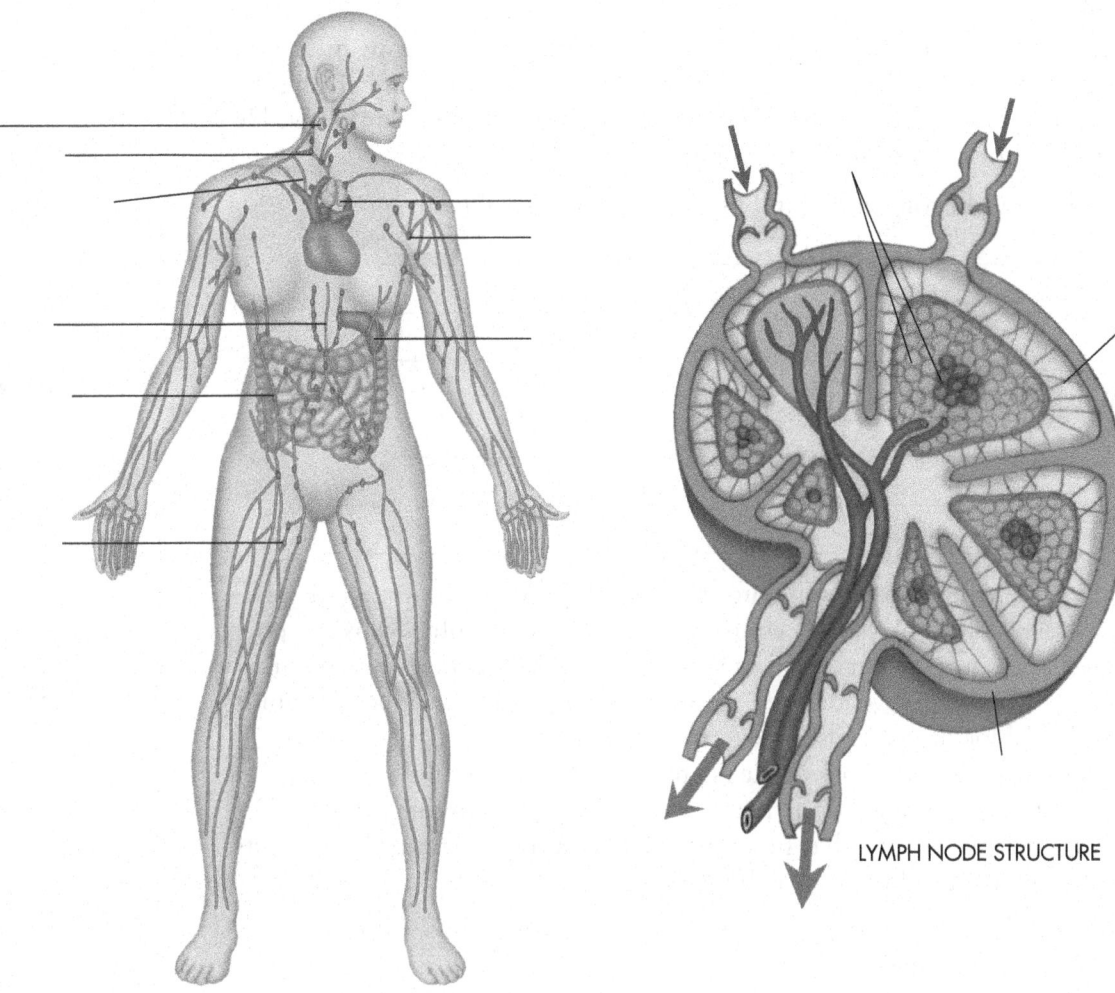

LYMPH NODE STRUCTURE

CASE STUDY

Jim, a hemophiliac, has been feeling weak and sick for several weeks, as well as fatigued, feverish, and he is losing weight. His lymph nodes are swollen, and he has a runny nose and sometimes a sore throat and cough. He doesn't really feel bad enough to stay home from work, but wishes he could just stay in bed some mornings. At first he thought it was his allergies acting up or a sinus infection, but antihistamines and decongestants haven't helped. Then he thought maybe he had a virus, but he has been sick for too long and isn't feeling any better. He goes to the doctor, who runs a battery of tests.

1. Given Jim's history and symptoms what are the possible diagnoses?

     _____

     _____

     _____

2. What tests should be run?

     _____

     _____

     _____

Three days later, Jim's test results are in. His HIV test is negative. His CBC reveals elevated white blood cell count, particularly neutrophils. His lymphocyte numbers are near normal, and there is no sign of abnormal cells. A throat culture comes back positive for strep throat, and his tonsils are enlarged. Jim is prescribed antibiotics and sent home to recuperate.

  Two months later he is back at the doctor with the same symptoms. Another round of antibiotics follows and then another, but his symptoms return. His tonsils are now so enlarged that he has trouble sleeping. Repeated blood tests rule out leukemia, lymphoma, and HIV.

3. What did the doctor eventually suggest as a treatment to solve Jim's repeated infections?

     _____

     _____

     _____

LEARNING ACTIVITIES

1. Create a series of flashcards for each part of the immune system with the component on one side and the function on the other.

2. You are the immune system. Flu viruses have invaded your body. What do you do? List the steps taken by the immune system to fight off the virus.

3. Patients with suppressed immune systems are often victims of opportunistic infections. What is an opportunistic infection? List the common opportunistic infections found in patients who are immune-compromised.

4. Play immune system "Jeopardy."

5. For each part of the immune system, come up with an analogy not found in the book. The book used warfare and fire departments. Can you think of other analogies?

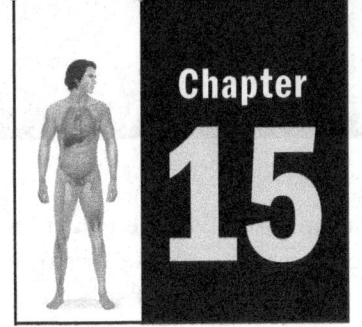

THE GASTROINTESTINAL SYSTEM: FUEL FOR THE TRIP

Chapter 15

CHAPTER SUMMARY

The digestive system is responsible for the ingestion, digestion, absorption of nutrients, and elimination of waste from the food. It can be divided into the alimentary canal, the actual pipes through which food travels, and a series of accessory organs that secrete substances that aid digestion.

The alimentary canal begins with the oral cavity or mouth. The job of the mouth is to moisten food, begin mechanical breakdown of food, taste food, and begin digestion of starches. The teeth and tongue aid in mechanical breakdown, and the salivary glands, accessory glands that open into the mouth, secrete saliva, which moistens food and begins starch digestion. Disorders of the oral cavity generally center on tooth health, including cavities and periodontal disease.

After leaving the mouth, food enters the pharynx, which was discussed in detail in Chapter 13. From the pharynx, the food enters the esophagus, a muscular tube that transports food to the stomach. No digestion takes place in the esophagus or the pharynx. The swallowing reflex, which protects the trachea when food is passing through the pharynx, is mediated by the pharynx. There are sphincters on each end of the esophagus: the pharyngoesophageal sphincter on the superior end and the lower esophageal sphincter on the inferior end at the opening to the stomach. Food moves down the esophagus on waves of contraction known as peristalsis.

The walls of the alimentary canal are variations on a theme, with each organ having the same general structure with minor changes related mainly to their functions. The inner layer of the wall is a mucosa. The next layer is the submucosa, a connective tissue layer with blood vessels, lymphatic tissue, nerves, and glands. The next layer out is smooth muscle, a layer of circular (changes tube diameter) and a layer of longitudinal (changes tube length) muscle. The outer layer is the serosa, typically the visceral peritoneum. The peritoneum is a serous membrane in the abdominal cavities. Like the pericardium and the pleura, the peritoneum has a parietal layer that lines the abdominal cavity and a visceral layer that covers the organs. Between the layers is a potential fluid-filled space, the peritoneal cavity. Unlike in the thoracic cavity, some abdominal organs are not covered by visceral peritoneum but are retroperitoneal.

The stomach is a J-shaped pouch hanging below the diaphragm. Food enters the stomach from the esophagus into an area known as the cardiac region. The dome-shaped fundus is just superior to the cardiac area. The major part of the stomach is the body, with the greater and lesser curvatures. Food leaves the stomach to enter the small intestine via the pyloric region. The stomach functions as a storage chamber and as a digestive organ, beginning the digestion of proteins. It also absorbs a small number of substances, including alcohol. The stomach has several unique anatomical features. The mucosa and submucosa are folded into ridges called rugae that allow the stomach to expand as it fills. There is an extra layer of smooth muscle, an oblique layer, that allows the stomach to make complex mixing movements. The stomach secretes a number of substances, including hydrochloric acid, mucus, and pepsin, a protein-digesting enzyme. This combination of secretions is known as gastric juice.

The activity of the stomach is controlled by the vagus nerve (cranial nerve X) and has three phases. The cephalic phase occurs when you see, smell, or taste food. This sensory stimulation stimulates the vagus nerve via the medulla oblongata. The vagus nerve stimulates the stomach, and the stomach releases a hormone, gastrin, which causes gastric juice to be released. The next phase of gastric activity, the gastric phase, begins when food enters the stomach. As food enters the stomach, the stomach stretches, stimulating receptors in the stomach wall. This information is relayed to the brain stem and via the parasympathetic nervous system signals are sent to the stomach to increase gastric juice secretion and muscle contraction. Food is moved into the small intestine. As soon as food enters the small intestine, the third phase of gastric secretion, the intestinal phase, begins. Stimulation of the small intestine by the entry of food causes hormones to be released from the small intestine that slow down stomach activity, thus controlling the movement of food into the small intestine.

As you might imagine, the presence of strong acid in the stomach is involved in stomach pathology. Gastroesophageal reflux disease (GERD) is a common disorder that occurs when excess stomach acids moves up into the esophagus, irritating it and causing the familiar sensation of "heartburn." Untreated GERD can cause long-term problems due to irritation of the esophagus. Antacids and acid-reducing medications can treat symptoms, but only lifestyle changes can prevent GERD. Peptic ulcers are the result of breakdown of the mucosa. In the past, ulcers were though to be caused by stress. However, we now understand that many ulcers are caused by a specific bacterial infection or weakness in the mucosa.

From the stomach, food enters the small intestine through the pyloric sphincter. The small intestine (small in diameter, not in length) is divided into three parts. The initial short section that receives food and secretions is the duodenum, the next portion is the jejunum, and the last portion is the ileum. Almost all the digestion of food and absorption of nutrients in your digestive system takes place in the small intestine. It has several adaptations for increasing surface area including villi (fingerlike projections of the mucosa and submucosa), plicae circularae, and microvilli (fingerlike projections on the surface of the mucosal cells). The small intestine secretes enzymes to digest food and hormones to control the activity of the stomach and accessory organs. Food moves from the small intestine into the large intestine through the ileocecal sphincter.

The large intestine is very simple anatomically and physiologically compared to the small intestine. The functions of the large intestine are limited to water reabsorption, absorption of vitamins produced by natural bacterial community housed in the organ, and production of feces. Digested food enters the cecum, a blind-end pouch to which the appendix is attached. From the cecum, food passes into the ascending, transverse, descending, and sigmoid colons before entering the rectum and passing out of the body via the anus. A pair of sphincters allows voluntary control of defecation. The large intestine has no villi nor microvilli and has reduced smooth muscle. Disorders of the large intestine include diverticulitis, colorectal cancer, and hemorrhoids.

Three accessory organs are important in the function of the digestive system: the liver, the gallbladder, and the pancreas. The liver is a large organ located in the upper right portion of the abdomen. It is divided into four lobes. The liver is very well vascularized. Blood from the digestive organs drains into the liver via the hepatic portal vein, and oxygenated blood enters via the hepatic vein. The liver has many functions, including glycogen storage, making cholesterol, detoxifying blood, making proteins, iron storage, storage of fat-soluble vitamins, and making urea. The liver is directly involved in digestion because it produces bile, an important emulsifier. Bile is made in the liver and released into the duodenum when the duodenum releases the hormone secretin. Bile can be stored in the gallbladder. The gallbladder is a pear-shaped storage sac nestled under the liver. It stores bile produced by the liver and releases it into the duodenum under the influence of the hormone CCK, released by the duodenum.

The pancreas is a gland that secretes digestive enzymes into the duodenum. (It also is an endocrine gland that secretes insulin and glucagon into the blood to control blood sugar.) The pancreas secretes digestive enzymes into the duodenum under the influence of the hormone CCK, released by the duodenum. All these accessory organs can become inflamed due to infection or injury. Cholecystitis, inflammation of the gallbladder, is generally caused by obstruction of the cystic duct (the bile duct leading from the gallbladder) by gall stones. Hepatitis, inflammation of the liver, is usually caused by a viral infection, toxins, or certain drugs. Pancreatitis can be caused by a variety of irritants, including alcoholism and cholecystitis. Pancreatitis is very painful and if untreated can be fatal. There are many other digestive system disorders, but most have similar symptoms, including vomiting, diarrhea or constipation, and pain. See the table at the end of the chapter for a list of disorders.

CHAPTER OUTLINE

MEDICAL TERMINOLOGY REVIEW

Define the following terms.

1. Cholecystitis: _____

2. Pancreatitis: _____

3. Hepatitis: _____

4. Diverticulosis: _____

5. GERD: _____

6. Caries: _____

7. Cirrhosis: _____

8. Gastroenteritis: _____

9. Peptic ulcer: _____

10. Peritonitis: _____

MULTIPLE CHOICE

Circle the letter of the correct answer.

1. Which of the salivary glands is found on the roof of the mouth?
   a. Parotid
   b. Submandibular
   c. Sublingual
   d. None of the above

2. If not pulled or knocked out, how many permanent teeth do we have by 25 years of age?
   a. 32
   b. 16
   c. 28
   d. 42

3. Please arrange the segments of the large intestine in the order waste travels through.
   a. Cecum, descending colon, ascending colon, transverse colon, sigmoid colon, rectum
   b. Cecum, ascending colon, transverse colon, descending colon, sigmoid colon, rectum
   c. Sigmoid colon, ascending colon, transverse colon, descending colon, cecum, rectum
   d. Sigmoid colon, descending colon, ascending colon, transverse colon, cecum, rectum

4. When food enters the mouth it is said to be:
   a. ingested.
   b. digested.
   c. absorbed.
   d. All of the above

5. The digestive enzyme secreted by the pancreas that when activated in the small intestine will digest proteins in food is called:
   a. peptidase.
   b. amylase.
   c. pepsin.
   d. cholecystokinin.

6. Where does 80% of absorption of usable nutrients take place?
   a. Stomach
   b. Mouth
   c. Large intestine
   d. Small intestine

7. Which nerve innervates the visceral muscles of the stomach, causing contraction and hence motility?
   a. Phrenic
   b. Vagus
   c. Trigeminal
   d. Sciatic

8. The labia is/are commonly known as the:
   a. tongue.
   b. gallbladder.
   c. lips.
   d. uvula.

9. What does emulsify mean in terms of fat?
   a. The building of fatty acid chains in the liver
   b. The binding of fatty acids to carrier proteins for transport to the liver
   c. The destruction of fat globules or the rendering of fat globules unusable so no absorption will ever take place
   d. The breaking or converting of fat into a form that promotes enzymatic chemical digestion

10. Which sphincter lies between the stomach and small intestine?
    a. Cardiac
    b. Gastroenteral
    c. Pyloric
    d. Ilieocecal

11. What is the pH (acidity) of HCL in the stomach?
    a. 1.5 to 2.0
    b. 7.0 to 7.2
    c. 7.5 to 8.8
    d. 12.0 to 13.6

12. What is the function of the liver?
    a. Detoxify
    b. Produce clotting factors
    c. Store glucose in a form called glycogen
    d. All of the above

13. What is a lacteal, and where is it located?
    a. Lymphatic capillary in each villus of small intestine
    b. Blood capillary beside goblet cells in the pancreas
    c. Enzyme in the pancreas that, when secreted, digests milk
    d. Mucous lining found in the stomach

14. In the stomach, what do the parietal cells secrete, and what do the chief cells secrete?
    a. Sucrose/fructose
    b. Amylase/lipase
    c. HCL/pepsinogen
    d. Bile/bilirubin

15. In reference to the cardiac sphincter, where is the fundus of the stomach?
    a. Left, superior
    b. Right, inferior
    c. Left, inferior
    d. Right, superior

16. What structure prevents us from swallowing our tongue and also aids in proper speaking?
    a. Diaphragm
    b. Uvula
    c. Epiglottis
    d. Frenulum

17. Which section of the small intestine connects to or is continuous with the stomach?
    a. Cecum
    b. Duodenum
    c. Ileum
    d. Jejunum

18. Which of the following statements is correct?
    a. The hepatic ducts conduct bile from the liver, the cystic duct conducts bile to and from the gallbladder, and the common bile duct conducts bile to the small intestine.
    b. The common bile duct conducts bile from the liver, the cystic duct conducts bile to and from the gallbladder, and the hepatic ducts conduct bile to the small intestine.
    c. The cystic duct conducts bile from the liver, the hepatic ducts conduct bile to and from the gallbladder, and the common bile duct conducts bile to the small intestine.
    d. The common bile duct conducts bile from the liver, the hepatic ducts conduct bile to and from the gallbladder, and the cystic ducts conduct bile to the small intestine.

19. Where is the most common region for peptic ulcer disease?
    a. Distal and middle parts of the esophagus
    b. Body of the stomach
    c. Upper or proximal part of small intestine
    d. Rectum and around the anal sphincter

20. The vermiform appendix hangs off the:
    a. Cecum
    b. Rectum
    c. Colon
    d. Ileum

21. What effect does secretin have on the stomach?
    a. Increases muscular activity
    b. Produces bile
    c. Increases secretions
    d. Decreases overall activity

22. How many incisors do adults normally have?
    a. 4
    b. 6
    c. 8
    d. 10

23. The uvula is associated with which structure?
    a. Soft palate
    b. Hard palate
    c. Tongue
    d. Pharynx

24. What substance starts chemically breaking down in the mouth due to salivary secretions?
    a. Starch
    b. Protein
    c. Fat
    d. Lactose

25. Bilirubin from what is eliminated in bile?
    a. Fat
    b. Food
    c. Feces
    d. Blood cells

26. _____ is/are the result of microorganisms attacking tooth enamel.
    a. Periodontal disease
    b. Caries
    c. Gingivitis
    d. Leukoplakia

27. The abbreviation GERD stands for:
    a. gastric enzyme reduction disease.
    b. gastric electrical reflex disease.
    c. gastroesophageal reflex disorder.
    d. gastroesophageal reflux disease.

28. Which of the following is not a risk for colorectal cancer?
    a. Diet rich in animal fat
    b. Diet high in fiber
    c. Sedentary lifestyle
    d. High blood cholesterol

29. Scarring of the liver is known as:
    a. cirrhosis.
    b. hepatitis.
    c. jaundice.
    d. HBV.

30. This digestive disorder can be treated with a low gluten diet:
    a. gastroenteritis.
    b. diverticulitis.
    c. GERD.
    d. None of the above

31. This condition may be a precursor to oral cancer:
    a. polyposis.
    b. leukoplakia.
    c. intussusception.
    d. GERD.

32. Which of the following disorders has heartburn as a symptom?
    a. GERD
    b. Reflux esophagitis
    c. Hiatal hernia
    d. All of the above

33. This is the technical term for "failure to thrive":
    a. malabsorption syndrome.
    b. anorexia nervosa.
    c. bulimia nervosa.
    d. irritable bowel syndrome.

34. This disorder could be a result of injury to the large intestine that allows intestinal bacteria to escape into the abdominal cavity:
    a. peritonitis.
    b. appendicitis.
    c. diverticulitis.
    d. hepatitis.

35. Pain in the lower right quadrant of the abdomen is a symptom of:
    a. hepatitis.
    b. peritonitis.
    c. pancreatitis.
    d. appendicitis.

MATCHING EXERCISES

Set 1

Please match each term with the appropriate definition.

_____ 1. Cirrhosis
_____ 2. Enteritis
_____ 3. Anorexia
_____ 4. Calculi
_____ 5. Hemorrhoids
_____ 6. Crohn's disease
_____ 7. Gingivitis
_____ 8. Gastritis
_____ 9. Bulimia
_____ 10. Cholecystitis

a. Disease marked by "binge-purge" of food
b. Inflammation of the stomach
c. Chronic disease of the liver
d. Inflammation of the gallbladder
e. Constipation or fecal impacting at the transverse colon
f. Regional ileitis
g. Inflammation of the small intestine
h. Disease marked by loss of appetite and remarkable weight loss
i. Inflammation of the gums
j. Varicose veins of the rectum
k. Gallstones

Set 2

Please match each chemical with the appropriate function.

_____ 1. Secretin
_____ 2. Cholecystokinin
_____ 3. Pepsin
_____ 4. Hydrochloric acid
_____ 5. Bile
_____ 6. Peptidase
_____ 7. Intrinsic factor
_____ 8. Gastrin
_____ 9. Sucrase
_____ 10. Amylase

a. Breaks down protein in stomach
b. Emulsifies fat
c. Breaks down starches in mouth
d. Neutralizes the chyme in duodenum
e. Needed for the absorption of B_{12}
f. Stimulates the release of bile
g. Breaks down dissacharides
h. Breaks down protein in small intestines
i. Hormone that increases gastric activity
j. Digests portions of protein structures in small intestine
k. Converts pepsinogen to pepsin
l. Hormone that activates bile production

Set 3

Please match each term with the appropriate definition.

_____	1.	Chyle
_____	2.	Rugae
_____	3.	Gingiva
_____	4.	Peyer's patch
_____	5.	Adventitia
_____	6.	Villi
_____	7.	Epiglottis
_____	8.	Nitroglycerine
_____	9.	Cementum
_____	10.	Bolus

a. Medication used to increase gastric juices
b. Folds in the stomach
c. Anchors root of tooth to gums
d. Lymph tissue in small intestine
e. Food stuff mixed with salivary juices
f. Food stuff mixed with gastric juices
g. Fingerlike protrusions in small intestine
h. Absorbable under the tongue
i. Gum
j. Outer layer of the esophagus
k. Prevents food from slipping into lungs
l. Lipoproteins in the lacteal formed from glycerol and fatty acids

Set 4

Please match each disorder with the appropriate treatment.

_____	1.	Peritonitis
_____	2.	Irritable bowel syndrome
_____	3.	Diverticulitis
_____	4.	Crohn's disease
_____	5.	GERD
_____	6.	Caries
_____	7.	Anorexia nervosa
_____	8.	Leukoplakia
_____	9.	Polyposis
_____	10.	Cholelithiasis

a. Decrease stomach acid, avoid trigger food, elevate head while sleeping
b. Cease tobacco use
c. High fiber diet, stool softener, antibiotics, surgery
d. Correct cause, surgery, antibiotics
e. Fillings
f. Lifestyle changes, diet, medication
g. Surgery
h. Counseling, better nutrition
i. Anti-inflammatory medications, surgery
j. Dietary changes, observation, gallbladder removal

FILL IN THE BLANK

Fill in the blanks to complete the following statements.

1. Any organ, the function and size of which seem to have been reduced as humans evolved, is termed

   _____.

2. Another name for canine teeth is

   _____.

3. Gastric activities, such as churning and secretion of enzymes, are controlled by the

   _____ nervous system (be specific).

4. For most of the digestive tract, the serosa layer is also called
    _____.

5. Between the ages of 2 and 3, all
    _____ of your baby teeth
    should have appeared.

6. The clinical term for the elimination of unusable material from the body
    is _____.

7. The digestive tract is also called the
    _____ tract.

8. If fecal material moves through the large intestine too fast,
    _____ occurs.

9. Now violently coughing, client A was previously eating and talking at the
    same time. Her partially chewed beef jerky slipped by her
    _____, which closes off
    her airway when she swallows.

10. The three main regions of the large intestine are the
    _____,
    _____, and
    _____.

11. Besides food from the stomach, the first part of the small intestine
    receives additional secretions from the
    _____ and the
    _____.

12. Baby teeth are clinically called
    _____ teeth.

13. Heartburn occurs when the
    _____ opens and there
    is a backflow of food.

14. The digestive enzyme found in saliva is
    _____.

15. If fecal material moves too slowly through the intestine,
    _____ occurs.

16. Bypassing of part of the large intestine is called a(n)
    _____.

17. _____ are mouth ulcers
    sometimes caused by eating certain foods.

18. A(n) _____
    tube allows food to be directly injected into the stomach.

19. People who cannot digest dairy products are often
    _____ intolerant.

20. The treatment for appendicitis is
    _____, removal of the
    appendix.

21. _____ are varicose veins in
    the anus.

22. Exposure to ingested toxins can cause damage to this organ:
    _____.

23. Inflammation of the gallbladder may lead to inflammation of this nearby accessory organ:
    _____.

24. Health care workers and first responders often get immunized for the
    _____ virus.

25. _____, a yellowish tinge to the skin, is one symptom of liver disease.

SHORT ANSWER

1. What is the purpose of villi, plicae circularis, and microvilli in the small intestine?

    _____

    _____

    _____

    _____

2. Why do lactose-intolerant people suffer from diarrhea and bloating if they consume milk or milk products?

    _____

    _____

    _____

    _____

3. Although it produces powerful enzymes and chemicals, why doesn't the stomach digest itself?

    _____

    _____

    _____

    _____

4. List and describe the location of the three named salivary glands.

    _____

    _____

    _____

    _____

5. What is the importance of digestion?

    _____

    _____

    _____

    _____

6. Why are digestive disorders sometimes difficult to diagnose?

   _____

   _____

   _____

   _____

7. Several digestive disorders may be improved by lifestyle changes.
   List them.

   _____

   _____

   _____

   _____

LABELING ACTIVITIES

1. Label the organs and structures in the figure. Color code each structure using Figure 15–1 of your textbook as a guide.

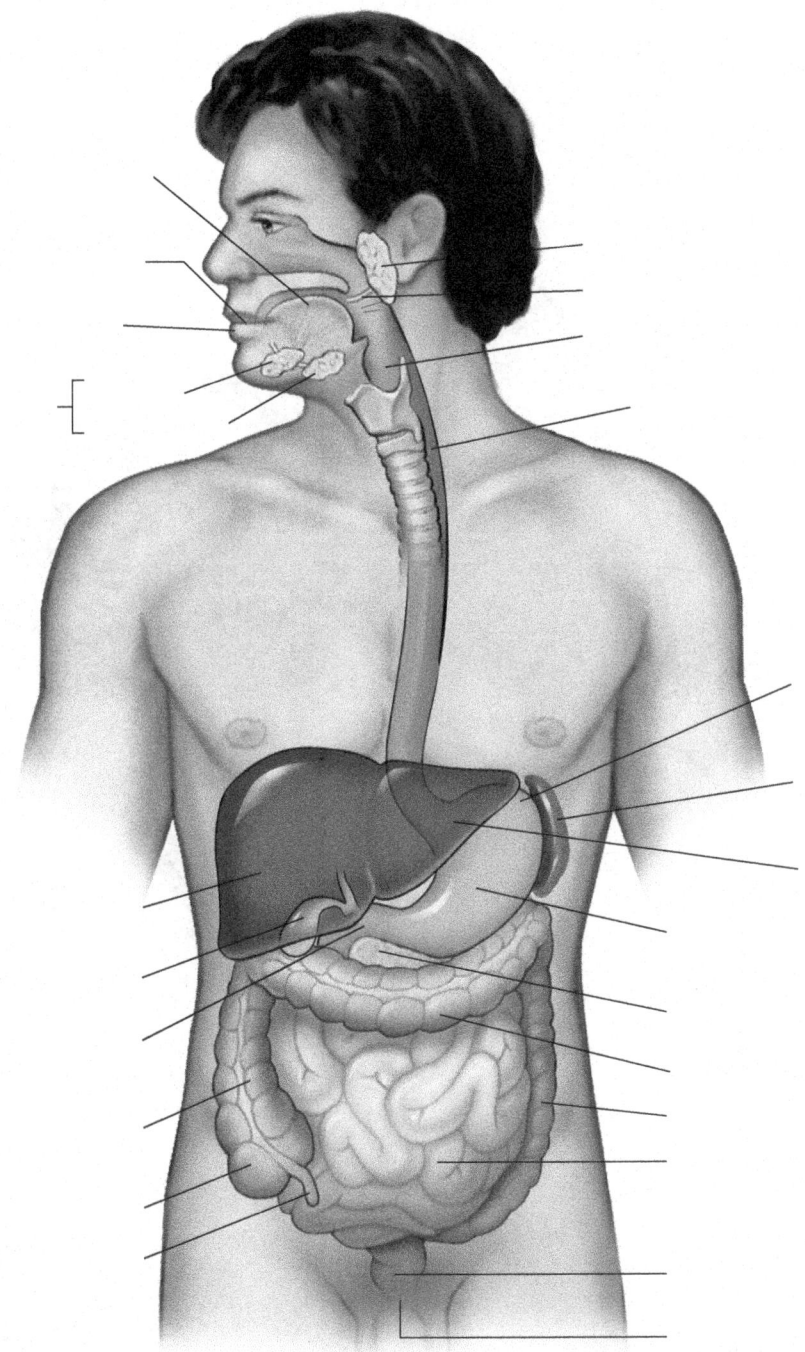

2. Label the structures in the wall of the alimentary canal using Figure 15-6 of the text.

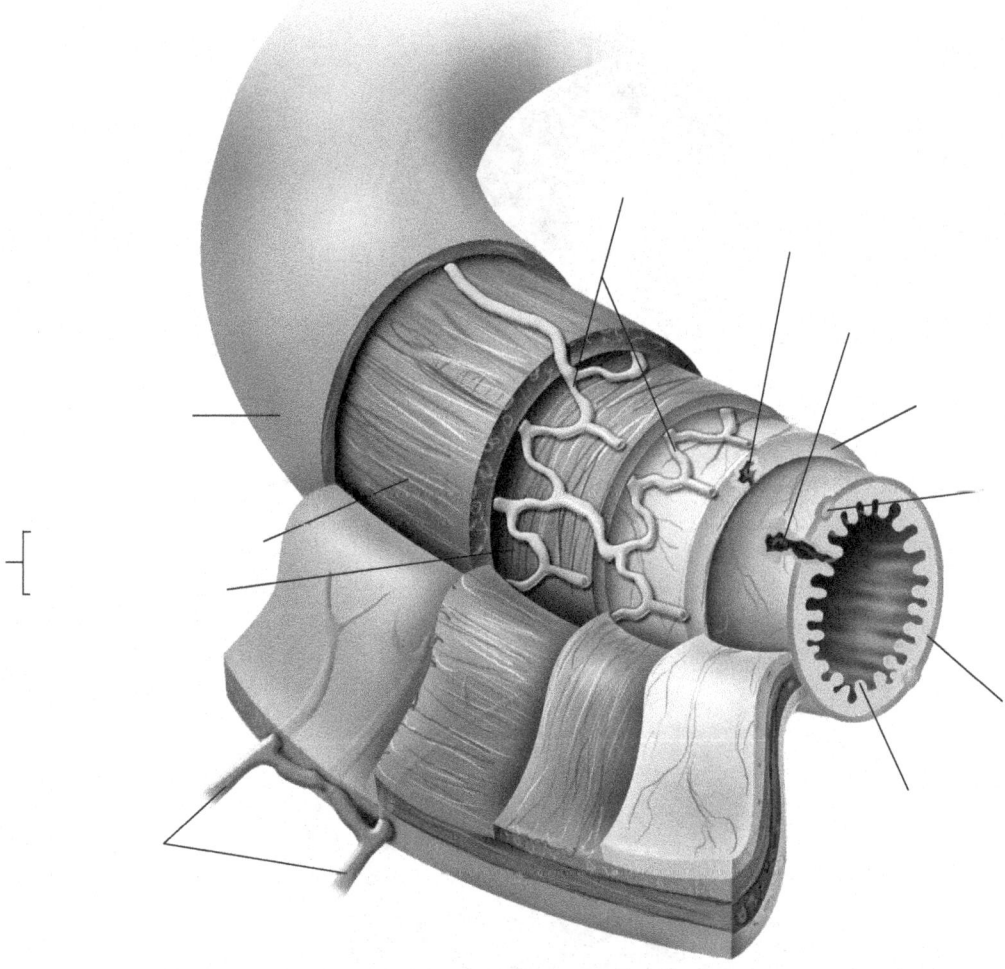

CASE STUDY

Vera, a 25-year-old schoolteacher, wakes up one morning freezing cold, with terrible abdominal pain. The pain is so bad she is not sure she can get out of bed. She had felt vaguely ill for several days, but now there is no doubt that something is wrong. As she struggles out of bed, the pain is joined by nausea. Staggering to the bathroom, she is violently ill several times. Wondering why she is so cold, she takes her temperature and is shocked to find it is 103°F. Truly alarmed, she calls in sick for work, calls her physician (who tells her to head for the emergency department), and calls a friend to take her to the hospital.

1. Why does Vera's physician suggest she go to the emergency department instead of waiting for an appointment later in the day?

    _____

    _____

    _____

2. What possible disorders do Vera's symptoms suggest?

    _____

    _____

    _____

When the ER personnel take Vera's history, she tells them she had her appendix out when she was 10 years old. She has no history of digestive disorders, and she is a competitive athlete and a vegetarian.

3. What disorders can be ruled out by Vera's history and lifestyle?

    _____

    _____

    _____

Vera's blood work shows her white blood cells are extremely high, suggesting a massive infection. Upon further questioning Vera tells the physicians that she recently had a bicycle accident, flying over the handlebars of her mountain bike and onto a rock pile. Stiff and bruised, but not incapacitated, she hadn't given it another thought. A CT scan reveals that Vera has a small tear in her large intestine which, untreated, has lead to peritonitis and borderline septicemia.

4. What is the treatment for Vera's problem?

    _____

    _____

    _____

Vera underwent treatment, spent several days in the hospital and recovered fully.

LEARNING ACTIVITIES

1. For each part of the digestive system, list the function.

2. Design a digestive system board game. Draw the alimentary canal and accessory organs. Each space should be associated with a question about a specific part of the digestive system. Each player is food. The object is to get from the plate to the anus. Roll dice to determine how many spaces can be moved. Answering the question correctly allows movement on the board.

3. For each of the major digestive disorders list the symptoms. How similar are they? How can you make an accurate diagnosis?

4. Diet has a great deal of influence on digestive health. List the kinds of foods that often cause trouble. Which disorders are obviously influenced by food choices?

5. Colorectal cancer is a common cancer in the United States. Do some research on colorectal cancer. What is the screening test? Why are polyps important? How can colorectal cancer be prevented? How can it be treated?

THE URINARY SYSTEM: FILTRATION AND FLUID BALANCE

CHAPTER SUMMARY

The job of the urinary system is to control fluid volume and electrolyte balance by controlling urination. The system consists of paired kidneys and ureters and a single urinary bladder and urethra. To make urine, the nephron, the functional unit of the kidney, must perform three processes: filtration, reabsorption, and secretion.

The kidneys are paired organs in the superior dorsal part of the abdominal cavity. They are 10 cm long, bean shaped, and covered by a capsule. The ureter, renal artery, and renal vein are attached at an indentation called the renal hilum. Internally the kidney can be divided into three regions. The outer region, the renal cortex, is grainy in appearance. The middle region is the renal medulla, and the innermost region is the renal pelvis. The renal medulla has several structures called renal pyramids, triangular striped regions, separated by unstriped renal columns. The renal pelvis is a large funnel divided into large pipes known as major calyces. The major calyces are divided into smaller pipes known as minor calyces. Urine is filtered in the renal cortex, transported in the renal pyramids, and collected in the renal pelvis, which opens into the ureter. Urine then flows down the ureter to the bladder.

Because the kidney is filtering urine from blood, it must be well vascularized. Each kidney has a renal vein coming off the abdominal aorta. The renal artery branches several times into smaller and smaller arteries and arterioles until there are millions of microscopic capillaries in the renal cortex. Blood then flows back out through a series of venules and veins, leaving the kidney via the renal vein, which empties into the inferior vena cava. The columns in the renal medulla are passageways for blood vessels.

The fundamental functional unit of the kidney is called the nephron. Millions of microscopic nephrons make up each kidney. The nephron is divided into two parts: the renal corpuscle, which is a filter, and the renal tubules, in which reabsorption and secretion take place. The renal corpuscle has two parts: the glomerulus, which is a ball of capillaries, and the glomerular capsule, which receives the substance filtered from the blood in the glomerulus. The glomerular capsule is made of specialized epithelium, which acts as a filter. The substance filtered from blood into the glomerular capsule is known as glomerular filtrate.

After glomerular filtrate enters the glomerular capsule, it moves to the renal tubule. The renal tubule is actually several different tubules surrounded by capillaries called the peritubular capillaries and vasa recta. Substances can move between the capillaries and the tubules (reabsorption or secretion), controlling the chemistry and volume of urine. Filtrate flows from glomerular capsule to proximal tubule, descending nephron loop, ascending nephron loop, distal tubule, and into collecting ducts. The collecting ducts flow into the minor calyces, then to major calyces, and into the renal pelvis. The striped appearance of the renal pyramids is due in part to the collecting ducts running through the renal medulla. The epithelium making up the wall of the renal tubules is not the same in each tubule because each tubule does not have the same permeability. (More on this later.) Keep in mind the direction of movement of substances during urine formation. During filtration, substances move from blood into the glomerular capsule. During reabsorption, substances move from renal tubule into peritubular capillaries, and during secretion substances move from peritubular capillaries into renal tubule. What is in the renal tubule will eventually be part of urine and leave the body. Kidney function is dependent on clear passageways between the glomerulus and the renal pelvis. Kidney stones, crystals that block renal tubules, are very painful and may cause kidney damage if not treated. Polycystic kidney disease is a genetic condition in which large cysts develop in the kidneys, destroying nephrons. There is no treatment for PKD except kidney transplant. Decreased blood flow to the kidneys can also cause irreversible damage to kidney tissue.

Urine is formed by the kidneys as a mechanism for controlling fluid and ion balance and excreting nitrogenous waste. Three processes are necessary: filtration, reabsorption, and secretion. These processes are tightly regulated. Filtration is the movement of substances across a membrane from blood in the glomerular capillaries to glomerular filtrate in the glomerular capsule.

Two aspects of filtration are regulated: selectivity and rate. The selectivity of the filter is due mainly to the size of the holes in the filter. Proteins and blood cells are prevented from crossing the filter. Nearly everything else in blood passes through to the glomerular capsule. The chemistry of glomerular filtrate is identical to the chemistry of plasma except for large molecules like protein. Greatly increased pressure can also change filter selectivity. Filtration rate is controlled by pressure difference between one side of the filter and the other. It is controlled mainly by blood pressure (BP). When BP is within normal ranges, the kidney does not experience differences in filtration due to a mechanism called autoregulation. Autoregulation maintains steady pressure in the glomerulus even when systemic blood pressure is changing as long as BP is in normal range. However, in situations where BP must be regulated, filtration rate will change to help regulate BP. If systemic BP falls, the arterioles leading into the glomerulus constrict, decreasing blood flow to the filter and decreasing filtration rate. Ultimately less urine will be produced, raising fluid volume and thus raising BP. If systemic BP is too high, the opposite happens, and filtration increases. Constriction of the arterioles is accomplished by increased sympathetic output. Remember, sympathetic output decreases urination. The filtration apparatus can be damaged due to a condition called nephropathy. Nephropathy can be caused by diabetes mellitus, NSAIDs, antibiotics, immunosuppressants, and contrast dye used for imaging studies. Nephritis and glomerulonephritis, inflammation of the nephron, can cause permanent damage to the delicate filtration apparatus.

Tubular reabsorption and secretion control the chemistry of urine. Whereas glomerular filtrate is very much like plasma chemically, urine is very different. Some substances, like glucose, are completely reabsorbed in the renal tubules, whereas other substances, like urea, are secreted in large amounts. There are many regulators of tubular reabsorption and secretion. Tubular permeability is the first regulator. The proximal tubule is permeable to many different molecules, whereas all the other portions of the renal tubule are permeable only to water, ions, or both. Thus many molecules like glucose, antibiotics, and amino acids can only be reabsorbed or secreted in the proximal tubule.

The nephron loop is set up for countercurrent circulation. The descending loop is permeable to water and urea, whereas the ascending loop is permeable to ions and urea. There is a concentration gradient in the fluid surrounding the nephron loop such that water is reabsorbed in the descending loop and ions in the ascending loop, allowing the kidney to reabsorb both water and ions.

Tubular reabsorption and secretion are also regulated by several hormones that regulate blood pressure. Antidiuretic hormone decreases urine production by increasing reabsorption of water in the distal tubule and collecting ducts. Aldosterone, an adrenocorticosteroid, increases the reabsorption of sodium ions and excretion of potassium ions in the ascending loop and distal tubule. Aldosterone also decreases urination by increasing water reabsorption. Atrial natriuretic peptide is secreted by the atria when blood volume increases. It increases urination by decreasing sodium reabsorption. The renin-angiotensin-aldosterone system is a series of chemical reactions initiated by decreased blood flow to the kidneys. When blood flow to kidneys decreases, renin is released. In the bloodstream renin converts a molecule called angiotensinogen into angiotensin I. Angiotensin I is converted to angiotensin II in the lungs by angiotensin-converting enzyme. The effects of angiotensin II are increased aldosterone, increased thirst, and vasoconstriction. Urination decreases, and blood pressure rises.

Blood chemistry and blood pressure are dependent on adequate kidney function. When the kidneys do not work well, a condition known as uremia may develop. Uremia appears to be due to the accumulation of organic wastes in the bloodstream. It is a potentially fatal condition with widespread symptoms that are very difficult to treat. Hemolytic uremic syndrome is caused by kidney damage due to infection with *E. coli* bacteria. Renal failure is a collection of disorders ranging from short-term kidney malfunction to chronic, total renal failure. Damage to the kidneys from any number of causes can lead to renal failure. Acute renal failure is characterized by rapid decrease in kidney function, often caused by trauma, especially trauma resulting in blood loss and decreased blood flow to kidneys. Acute renal failure may be reversible or not. Chronic kidney disease is an ongoing, progressive kidney malfunction. The most common risk factor are diabetes mellitus and hypertension. Chronic kidney disease would eventually progress to end-stage renal failure, but most patients with kidney disease die from cardiovascular disease. The systems are intimately connected. Damage to one nearly always results in damage to the other. The only way to really prevent end-stage renal failure, or to treat it, is by kidney replacement, either dialysis, in which a machine performs the filtration functions of the kidney, or by kidney transplant. Each year thousands of patients die waiting for kidney transplants.

The kidneys make urine, but the urinary bladder stores and releases urine to the outside. Urine gets to the urinary bladder by flowing down paired ureters, one from each kidney. The bladder is a small, hollow, retroperitoneal organ posterior to the pubic symphysis. It is lined with transitional epithelium folded into rugae. The walls are smooth muscle. As urine accumulates, the smooth muscle stretches, which eventually triggers contractions of the bladder and urination. The reflex is mediated by neurons in the spinal cord and pons. However, voluntary control of urination, up to a point, is accomplished by a pair of urinary sphincters, which are made of skeletal muscle. Urine leaves the bladder via the urethra. There are several urinary bladder disorders. Incontinence is the inability to control urination. It has many underlying causes. Urinary tract infection is infection of the urinary bladder by fecal bacteria. Because of the anatomy of the urethra in women, urinary tract infection is much more common in women than in men. Overactive bladder is a common cause of incontinence. The cause of overactive bladder is not known but it is much more common in women and in people over the age of forty. For a list of urinary system disorders see the table at the end of the chapter.

CHAPTER OUTLINE

  I. Overview

 II. Kidney anatomy
      A. External
      B. Internal
           1. Gross anatomy
           2. Blood vessels
           3. Microscopic anatomy
                a. Nephron
                b. Pathology

III. Urine formation
      A. Chemistry
      B. Filtration
           1. Selectivity
           2. Filtration rate
           3. Pathology
      C. Tubular reabsorption and secretion
           1. Tubule permeability
           2. Hormones
           3. Pathology

 IV. Urinary bladder and urination

  V. Pathology

MEDICAL TERMINOLOGY REVIEW

Define the following terms.

1. Nocturia: _____

2. Enuresis: _____

3. Oliguria: _____

4. Pyuria: _____

5. Nephropathy: _____

6. Uremia: _____

7. Renal failure: _____

8. Kidney dialysis: _____

9. Creatinine: _____

10. Nephritis: _____

MULTIPLE CHOICE

Circle the letter of the correct answer.

1. The renal capsule covers the:
   a. kidney.
   b. glomerulus.
   c. bladder.
   d. afferent arteriole.

2. Which of the following will not pass through the glomerular epithelium into the nephron?
   a. RBC
   b. WBC
   c. Protein molecules
   d. All of the above

3. Glucose is _____ in glomerular filtrate than in urine:
   a. at the same concentration
   b. at a higher concentration
   c. at a lower concentration
   d. None of the above

4. The urinary bladder walls are composed of what type of muscle?
   a. Smooth
   b. Voluntary
   c. Skeletal
   d. Both b and c

5. One of the symptoms of kidney stones:
   a. pale urine.
   b. blood in urine.
   c. lower back numbness.
   d. excessive, uncontrollable, painless urination with continual expulsion of crystalline structures.

6. In plasma, urea is _____ than in urine:
   a. at the same concentration
   b. at a higher concentration
   c. at a lower concentration
   d. None of the above

7. Besides water, which of the following substances is usually found in urine at the bladder level?
   a. Glucose
   b. Ammonia
   c. Amino acids
   d. All of the above

8. Where are the kidneys located?
   a. Upper abdominal cavity
   b. Lower abdominal cavity
   c. Scrotal sac
   d. Pelvic cavity

9. Which of the urinary organs transports urine from the kidneys to the bladder?
    a. Nephrons
    b. Urethras
    c. Ureters
    d. Glomerulus

10. Urea and creatinine are _____ in urine than in glomerular filtrate.
    a. at the same concentration
    b. at a higher concentration
    c. at a lower concentration
    d. None of the above

11. Sodium is _____ in plasma than in urine.
    a. at the same concentration
    b. at a higher concentration
    c. at a lower concentration
    d. None of the above

12. In which region of the kidney is blood filtered?
    a. Pelvis
    b. Medulla
    c. Cortex
    d. Capsule

13. Which of the following structures is located in the renal medulla?
    a. Major calyces
    b. Pyramids
    c. Glomerulus
    d. Minor calyces

14. Normally, how can we consciously control the expulsion of urine from the body?
    a. Conscious control over the urinary bladder muscle
    b. Conscious control over the ureter sphincters
    c. Conscious control over the urethral sphincters
    d. Conscious control over the production of urine

15. Which of the following is the correct order in which blood arrives at the glomerulus?
    a. Renal artery, peritubular, arcuate, lobular, lobar, segmental, efferent arteriole
    b. Renal artery, arcuate, segmental, lobar, lobular, cortical radiate, afferent arteriole
    c. Renal artery, lobar, interlobar, lobular, cortical radiate, arcuate, efferent arteriole
    d. Renal artery, segmental, lobar, interlobar, arcuate, cortical radiate, afferent arteriole

16. Blood leaves the kidney's hilum via the:
    a. renal artery.
    b. inferior vena cava.
    c. efferent arteriole.
    d. renal vein.

17. As blood travels through the vessels that surround the nephrons, it exits the kidneys through a series of vessels that are in direct reverse of the arteries with one exception.
    a. There are no arcuate veins.
    b. There are no segmental veins.
    c. There are no lobular or lobar veins.
    d. There are extra veins called the juxtaglomedullary veins.

18. In plasma, sodium and potassium are _____ when compared to glomerular filtrate.
    a. at the same concentration
    b. at a higher concentration
    c. at a lower concentration
    d. None of the above

19. What happens at the Bowman's capsule?
    a. Excretion
    b. Secretion
    c. Filtration
    d. Reabsorption

20. Which of the following usually makes up a sizable portion of glomerular filtrate?
    a. Red blood cells and proteins
    b. White blood cells and proteins
    c. Water and glucose
    d. Ammonia and hydrogen ions

21. Which of the following is secreted at the nephron?
    a. Red blood cells and proteins
    b. White blood cells and sodium
    c. Water and glucose
    d. Ammonia and hydrogen ions

22. Glomerular filtrate flows from the renal corpuscle into the:
    a. loop of Henle.
    b. proximal convoluted tubules.
    c. distal convoluted tubules.
    d. collecting ducts.

23. Which of the following is either completely or partially reabsorbed, respectively, at the nephron?
    a. Red blood cells and potassium
    b. White blood cells and proteins
    c. Glucose and water
    d. Ammonia and hydrogen ions

24. Glomerular filtrate flows from the distal convoluted tubules into the:
    a. collecting ducts.
    b. ascending limb of the loop of Henle.
    c. descending limb of the loop of Henle.
    d. proximal convoluted tubules.

25. When systemic blood pressure has increased, what protective measures do the kidneys take?
    a. Vasodilate
    b. Vasoconstrict
    c. Shut down one kidney
    d. Nephrotic necrosis (spontaneous death of the nephrons)

26. _____ uses an endoscope and a small incision to remove large kidney stones.
    a. Extracorporeal shock wave lithotripsy
    b. Percutaneous nephrolithotomy
    c. Uteroscopy
    d. Cystoscopy

27. The most common cause of nephropathy is:
    a. diabetes mellitus.
    b. diabetes insipidus.
    c. contrast dye.
    d. ischemia.

28. Why does blood loss damage the kidneys?
    a. Increased BP damages the filter.
    b. Decreased blood flow causes tissue damage.
    c. Direct trauma to the kidneys often happens during an accident or trauma.
    d. Urine backs up into the kidney.

29. Uremia is caused by:
    a. increased ion concentration in blood.
    b. hormonal abnormalities.
    c. decreased blood volume.
    d. accumulation of organic wastes in blood.

30. Decreased urine output, uremia, and fluid retention are all symptoms of:
    a. early diabetic nephropathy.
    b. polycystic kidney disease.
    c. renal failure.
    d. overactive bladder.

31. Which of the following may be a cause of overactive bladder?
    a. Overactivity of stretch receptors
    b. Incontinence
    c. Urinary tract infection
    d. Kidney disease

32. Fever, abdominal pain, fatigue, bruising, and increased urination indicate this disorder:
    a. urinary tract infection.
    b. hemolytic uremic syndrome.
    c. nephritis.
    d. acute renal failure.

33. _____ is the most common genetic cause of kidney disease.
    a. Diabetes mellitus
    b. Glomerulonephritis
    c. Hemophillus influenza
    d. None of the above

34. Which of the following is a symptom of diabetes mellitus?
    a. Polyuria
    b. Oliguria
    c. Anuria
    d. Dysuria

35. Which urinary disorder can be confused with symptoms of endometriosis?
    a. Overactive bladder
    b. Interstitial cystitis
    c. Urinary tract infection
    d. Incontinence

MATCHING EXERCISES

Set 1

Please match each term with the appropriate description.

_____	1.	PKD
_____	2.	Analgesic nephropathy
_____	3.	Diabetes insipidus
_____	4.	Diabetes mellitus
_____	5.	Water toxicity
_____	6.	Glomerulosclerosis
_____	7.	Hemolytic uremic syndrome
_____	8.	Kidney stones
_____	9.	Hematuria
_____	10.	Urinary tract infection

a. Dangerously low blood sodium
b. Nephrons are replaced by cysts
c. Blood in the urine
d. Movement of fecal matter into urethra and bladder
e. Scarring of portions of the renal corpuscles
f. May be caused by overuse of ibuprofen
g. Insulin deficiency and hyperglycemia
h. RBC debris may block vessels to kidney
i. Too little antidiuretic hormone being produced and secreted
j. Can block kidney tubules

Set 2

Please match each term with the appropriate definition.

_____	1.	Rugae
_____	2.	Diffusion
_____	3.	Osmosis
_____	4.	Voiding
_____	5.	Filtration
_____	6.	Secretion
_____	7.	Reabsorption
_____	8.	Autoregulation
_____	9.	Vasoconstriction
_____	10.	Vasodilation

a. Increased blood vessel diameter
b. Decreased blood vessel diameter
c. Movement of ions and solutes from high to low concentration
d. Urination
e. Permit expansion of the urinary bladder
f. Movement of substances from tubules to capillaries
g. Controls blood pressure to nephrons
h. Movement of water from low ion to high ion concentration
i. Movement of blood substances from glomerulus into capsule
j. Movement of substances from capillaries to tubules

Set 3

Please match each structure with the appropriate description.

_____ 1. Ureter
_____ 2. Urethra
_____ 3. Kidney
_____ 4. Bladder
_____ 5. Nephron
_____ 6. Renal hilum
_____ 7. Renal pyramid
_____ 8. Minor calyces
_____ 9. Juxtaglomerular cells
_____ 10. Peritubular capillaries

a. Bean-shaped structure that filters blood and forms urine
b. Functional unit of the kidney
c. Striped areas in the renal medulla; collection of renal tubules
d. Transports urine from kidneys to bladder
e. Transports urine to outside the body
f. Wraps around nephrons; participates in secretion and reabsorption
g. Indentation on medial side of kidneys
h. Monitor blood flow to kidneys; secretes renin
i. Receive filtrate from collecting duct
j. Hollow holding structure for urine

Set 4

Please match each disorder with the appropriate treatment.

_____ 1. Glomerulonephritis
_____ 2. Urinary tract infection
_____ 3. Polycystic kidney disease
_____ 4. Kidney stones
_____ 5. Diabetic nephropathy
_____ 6. Hemolytic uremic syndrome
_____ 7. Overactive bladder
_____ 8. Renal failure
_____ 9. Uremia
_____ 10. Acute renal failure

a. Bladder training, sympathetic drugs
b. Antibiotics
c. Pain relief, uteroscopy, extracorporeal shockwave lithotripsy
d. Dialysis, difficult to treat, kidney transplant only cure
e. Treat underlying cause
f. Blood transfusion, kidney dialysis
g. Glycemic control, blood pressure meds, diet, kidney replacement
h. Blood pressure medications, treatment of underlying condition, prevention of cardio-vascular complications, kidney transplant or dialysis
i. Treat cause, replace fluid volume, normalize BP, treat infection, dialysis if needed
j. No cure, medication, kidney replacement

FILL IN THE BLANK

Fill in the blanks to complete the following statements.

1. The three processes necessary to regulate blood chemistry and make urine are _____, _____, and _____.

2. The vessels called _____ bring blood to the glomerulus, and the vessels called _____ leave the glomerulus with the unfiltered blood components.

3. The nephron is divided into two distinct parts called the _____ and the _____.

4. Beverages such as _____ and _____ may increase urination inappropriately.

5. The noninvasive treatment to break up kidney stones, called _____, involves shock waves.

6. The leading cause of kidney failure in the United States is _____.

7. The hormone that acts to retain more sodium in the body is _____.

8. High blood and urine glucose (sugar) is characteristic of a disorder called _____.

9. The three structures found at the renal hilum are the _____, _____, and _____.

10. The innermost region of the kidney is the _____.

11. The functional unit of the kidney is the _____.

12. Glomerular filtrate flows from the proximal convoluted tubules into the _____.

13. When molecules move from the capillary network into the distal convoluted tubules, that movement is termed _____.

14. The substances that are not filtered through the glomerular epithelium into the renal corpuscle and tubules are directed into the _____ vessels.

15. Contraction of the urinary bladder muscles is controlled by _____ neurons of the autonomic nervous system.

16. Dialysis in which blood is circulated through a machine and replaced is called _____.

17. Most patients with kidney disease die from

_____.

18. This substance, a waste product from muscle metabolism, is one indicator of kidney function

_____.

19. BUN measures

_____.

20. _____ in the urine is an early sign of nephropathy.

21. A treatment for early nephropathy is

_____. It may even slow progression.

22. Another name for urinary tract infection is

_____.

23. Hematuria is the presence of

_____ in the urine.

24. Ironically, _____, which must be taken after kidney transplant, are nephrotoxic.

25. Drugs that are the most effective at treating overactive bladder increase _____ activity.

SHORT ANSWER

1. Why are urinary tract infections more common in women than in men?

_____

_____

_____

_____

2. In maintaining homeostasis and in reference to the urinary system, how does autoregulation work?

_____

_____

_____

_____

3. How does the body regulate pH if too much acid is present in the blood?

_____

_____

_____

_____

4. In what way are aldosterone and atrial natriuretic hormones antagonists?

_____

_____

_____

_____

5. Why would a person who survives a trauma resulting in massive blood loss fall victim to kidney damage or permanent renal failure?

_____

_____

_____

_____

6. Explain the relationship between kidney disease and high blood pressure.

_____

_____

_____

_____

7. Explain the difference in diagnostic criteria between urinary tract infection and overactive bladder.

_____

_____

_____

_____

LABELING ACTIVITIES

1. Label the parts of the kidney using Figure 16–2 as your guide.

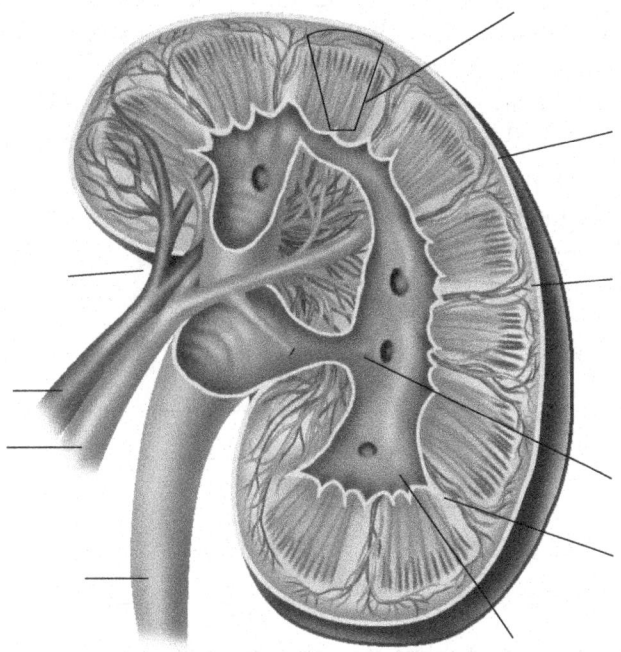

2. Label the parts of the nephron using Figure 16–4 as your guide.

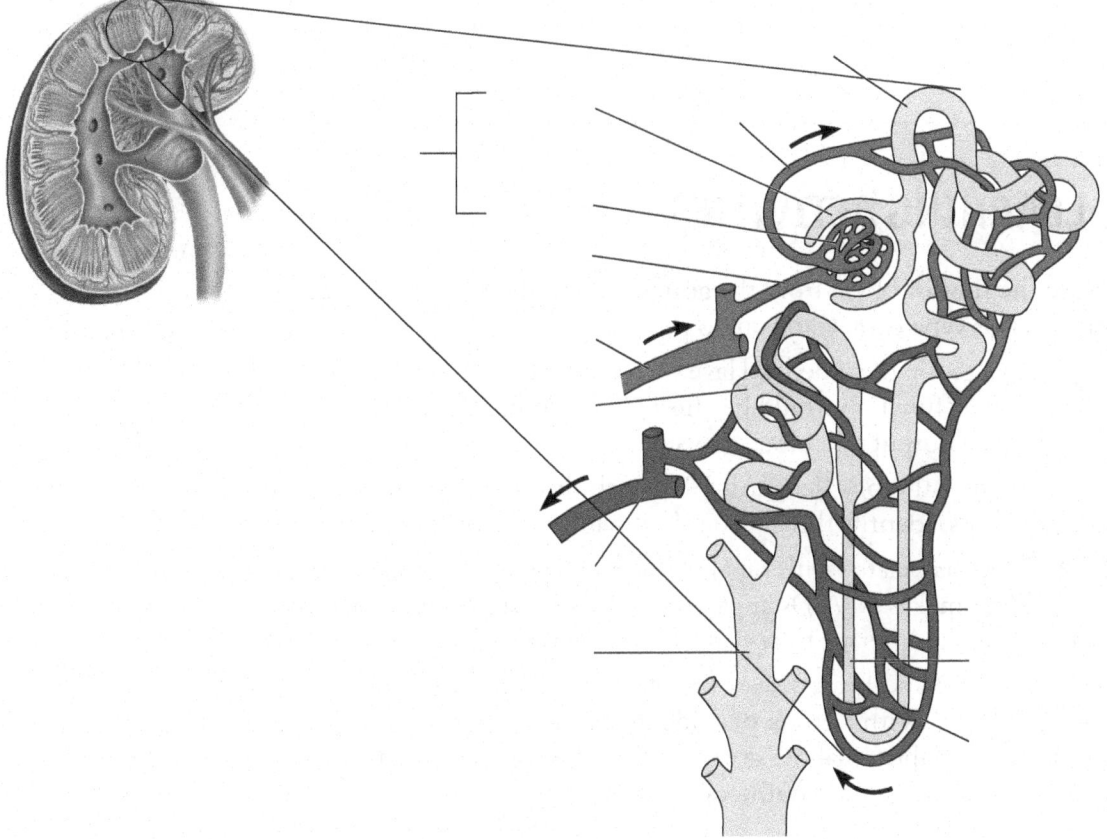

CASE STUDY

Barry, a diabetic, has been admitted to the hospital with an acute episode of mental confusion. In the ambulance he had a seizure. Assuming he is having an insulin reaction, causing dangerously low blood sugar, the hospital staff is shocked to find his blood glucose well above 200. They give him more insulin, but his blood glucose barely moves.

While waiting for the blood work, the physician talks to Barry's family and discovers that he has been feeling ill for a while. His symptoms include fatigue, lack of appetite, mental confusion, and pain in his feet. He has been having a terrible time controlling his blood sugar, so he had assumed that the symptoms were because of the poor glucose control.

He had made an appointment to see his doctor in a few days.

1. Given these symptoms, what will Barry's blood work show?

   _____

   _____

   _____

2. What will urinalysis show?

   _____

   _____

   _____

3. What, ultimately, is probably going on with Barry?

   _____

   _____

   _____

LEARNING ACTIVITIES

1. Trace the flow of blood from the aorta through the kidney and back to the inferior vena cava. Can you do it?

2. Organize a group of students. Have each student write a scenario in which kidney function would be affected. Try to predict what would happen to urine output and blood pressure.

3. Play urinary pathology "Jeopardy." For each disorder, list the symptoms or treatments. Identify the disorder in the form of a question.

4. NSAIDs are associated with renal failure in some individuals. Which ones? What makes NSAIDs apparently safe for some patients but not others? Use the Internet to research the question.

5. Kidney disease and heart disease are often found in the same patients. Why? With a group of students, brainstorm the reasons why cardiovascular abnormalities are often found in kidney patients. Use the Internet to check your hypotheses.

THE REPRODUCTIVE SYSTEM: REPLACEMENT AND REPAIR

 CHAPTER SUMMARY

Cells can reproduce either asexually or sexually. Asexual reproduction in human body cells is via mitosis. Mitosis is used for tissue growth, repair, and replacement. To make a new individual, humans must reproduce sexually, combining the DNA of two individuals.

To reproduce sexually, humans must produce gametes (eggs or sperm) with half the number of chromosomes as in their other body cells. Gametes are produced via meiosis, also known as reduction division. The human life cycle, then, has both mitosis and meiosis. Eggs and sperm are produced via meiosis in organs called gonads (testes in men, ovaries in women). The gametes get together and combine their DNA in a process known as fertilization. Fertilization results in a fertilized egg, also known as a zygote. The zygote undergoes many rounds of mitosis and development, eventually forming an embryo and then a fetus. The baby that results after a successful birth continues to grow and change via mitosis, eventually producing a sexually mature adult who can start the whole cycle over again.

The primary genitalia (the gonads) in females are the ovaries. The ovaries are in the pelvic cavity. They are covered by an epithelial capsule and are divided on the inside into a cortex, which contains the eggs, and medulla, which contains nerves and blood vessels. The uterine tubes (fallopian tubes, oviducts) are short tubes that carry eggs to the uterus and serve as a site for fertilization. The opening of the tube is surrounded by a large funnel (the infundibulum) with the ciliated fimbriae. The uterus is a fist-sized, thick-walled organ posterior to the urinary bladder. The uterus has a body, a fundus, and an isthmus. The cervix is the entrance to the uterus from the vagina. The wall of the uterus consists of three layers: the perimetrium, which is visceral peritoneum, the myometrium, a middle layer of muscle, and the endometrium, the mucosa. The endometrium has two layers: the functional layer, which is shed every month during menstruation, and the basal layer, which regenerates the functional layer every month. The uterus is well vascularized and supported by several ligaments and mesenteries. Sometimes bits of the endometrium implant in other parts of the body, especially in the abdominal and pelvic cavities, and shed each month, often causing severe pain and other complications. This condition is known as endometriosis.

The vagina is a smooth muscle tube that runs from the cervix to the outside. It receives the penis during intercourse, allows the flow of menstrual fluid out of the uterus, and serves as the birth canal.

The female external genitalia consist of the vaginal opening and structures that protect it, including the labia majora and minora. The clitoris is a small erectile structure. The breasts house the mammary glands, which secrete milk after pregnancy.

Female reproductive physiology is relatively complicated due to the hormonal changes that occur during the menstrual cycle. Women are born with all the eggs they will ever produce. The eggs stay dormant until a woman reaches puberty and her menstrual cycle begins with menarche, the first cycle. One egg will be released from the ovary (ovulation) approximately once a month until her last period, menopause. The cycle is divided into four stages, with the whole cycle lasting approximately 28 days. During the follicle stage, an egg matures under the influence of follicle-stimulating hormone. During ovulation, an egg is released under the influence of luteinizing hormone. During the corpus luteum stage, the ovary secretes the hormone progesterone to prepare the uterus for the potential arrival of a fertilized egg. If the egg isn't fertilized, the uterine lining is broken down and menstruation results. This is the final stage of the cycle. (If the egg is fertilized, it will implant in the endometrium, and pregnancy will result.) Once menses is over, the uterus begins to proliferate again, a follicle begins to develop in the ovary, and the cycle runs again. The cycle can also be broken down into two phases: the follicular or proliferative phase and the luteal or secretory phase. During the follicular phase, an egg is maturing and the endometrium is thickening. During the luteal phase, the corpus luteum is developing in the ovary and the tissue is beginning to break down in the uterus. The follicular phase precedes ovulation, and the luteal phase follows ovulation.

The cycle is tightly controlled by four hormones: estrogen and progesterone from the ovaries and follicle-stimulating hormone (FSH) and luteinizing hormone (LH) from the pituitary. FSH and LH levels rise under the influence of gonatotropin-releasing hormone from the hypothalamus. FSH initiates maturation and LH triggers ovulation. Prior to ovulation, a positive feedback loop keeps LH and FSH levels high and increases estrogen levels, which stimulates proliferation of the endometrium. Once ovulation occurs, the feedback loop becomes negative. Estrogen levels drop, and the ovary begins to secrete progesterone. GnRH, LH, and FSH levels also drop. This prevents another egg from maturing. After about 10 days, the ovary stops producing progesterone because the corpus luteum disintegrates. The decrease in progesterone leads to the shedding of the endometrium and increased levels of FSH, and LH. The cycle begins again. If pregnancy should occur, human chorionic gonadotropin will be released to prevent the shedding of the uterine lining. Many women report monthly symptoms that appear to be related to their monthly cycle. Mild symptoms are often classified as premenstrual syndrome, whereas debilitating symptoms are known as premenstrual dysphoric disorder.

The male primary genitalia are the testes, which are paired, egg-shaped organs suspended on either side of the penis in the scrotum. A testis is surrounded by an epithelial capsule and divided internally into a series of lobules filled with the seminiferous tubules. Sperm are born in the testes. The epididymis is a comma-shaped structure that sits on each testis like a hat. Sperm mature here. The penis is also part of the external genitalia. It is a sperm delivery organ.

The vas deferens is a long tube that transports sperm from the epididymis on the outside of the body into the pelvic cavity. The vas deferens joins the seminal vesicles posterior to the bladder to form the ejaculatory duct. The ejaculatory duct empties into the urethra to transport sperm to the outside via the penis. The seminal vesicles, prostate gland, and bulbourethral glands secrete substances into the sperm as it passes by, forming semen.

Spermatogenesis is much less complicated than oogenesis. Males do not begin to produce sperm until puberty, and they continue to produce sperm for the rest of their lives. Sperm production is controlled by testosterone, GnRH, LH, and FSH. At puberty males begin to secrete large amounts of testosterone, which enhances GnRH secretion and, therefore, LH and FSH. (Prior to puberty, the little testosterone secreted by the testes actually inhibits GnRH secretion.) LH and FSH stimulate sperm development.

When a man is aroused sexually, the penis becomes engorged with blood, expanding and stiffening. This is called erection. For sperm to be released, smooth muscle contractions throughout the system propel semen along the vas deferens, ejaculatory duct, and urethra and out the end of the penis. This is ejaculation.

Pregnancy results when a fertilized egg implants in the uterus. The period between implantation and birth should be about 40 weeks. From implantation until about week 8, the cells are called an embryo. After week 8 up until birth, it is called a fetus. The fetus is nourished by a connection to the mother called the placenta and surrounded by amniotic fluid. Labor is the actual process of birth and can be divided into three stages: dilation, expulsion, and placental.

CHAPTER OUTLINE

2. Oogenesis, follicle development, and ovulation
3. Hormonal control
4. Pathology
   C. Male anatomy
1. Testes
2. Penis
3. Epididymis
4. Vas deferens
5. Accessory glands
6. Pathology
   D. Male reproductive physiology
1. Spermatogenesis
2. Hormonal control
3. Erection and ejaculation
4. Pathology connection

V. Pregnancy
   A. Embryonic development
   B. Pathology

VI. Reproductive disorders

MEDICAL TERMINOLOGY REVIEW

Define the following terms.

1. Endometriosis: _____

2. Amenorrhea: _____

3. Ectopic pregnancy: _____

4. Placenta previa: _____

5. Erectile dysfunction: _____

6. Benign prostatic hypertrophy: _____

7. Hydrocele: _____

8. Androgen insensitivity: _____

9. Impotence: _____

10. Premenstrual syndrome: _____

MULTIPLE CHOICE

Circle the letter of the correct answer.

1. The external opening of the vagina may be covered by a perforated membrane called the:
   a. prepuce.
   b. foreskin.
   c. hymen.
   d. fimbria.

2. BPH is a pathology marked by:
   a. breast polyps commonly seen in women over 50.
   b. biological pelvic heliobacterium that affects prepubescent girls.
   c. prostate enlargement commonly seen in males over 50.
   d. hermaphroditic pelvic organs caused by fetal hormone imbalance.

3. In patients with erectile dysfunction disorder:
   a. the penis lacks blood vessels in the shafts.
   b. the penis is unable to lose an erection.
   c. men are unable to make viable sperm.
   d. the penis is not able to have a full erection.

4. Uterine tubes are not:
   a. the birth canals.
   b. the oviducts.
   c. the fallopian tubes.
   d. where fertilization takes place.

5. How many chromosomes does a zygote have in each cell?
   a. 42
   b. 46
   c. 21
   d. 16 pairs

6. The primary male genitalia is/are the:
   a. penis.
   b. testicles.
   c. sperm.
   d. muscles.

7. Which of the following are part of the spermatic cord?
   a. Vas deferens, ejaculatory duct, epididymis
   b. Testes, vas deferens, nerves
   c. Vas deferens, nerves, blood vessels
   d. Blood vessels, ejaculatory duct, penis

8. In most cases, vaginitis is caused by:
   a. microorganisms.
   b. macrooganisms.
   c. radiation.
   d. trauma.

9. In a pap smear, scrapings from what area are examined for precancerous cells?
   a. Cervix
   b. Vaginal walls
   c. Uterine walls
   d. Ovarian surface

10. Hydrocele is an abnormal collection of fluid in the:
    a. ovaries.
    b. breasts.
    c. testes.
    d. uterus.

11. What may cause amenorrhea?
    a. Emotional distress
    b. Extreme dieting
    c. Poor health
    d. All of the above

12. When a vasectomy is performed, what is prevented from traveling out of the penis during intercourse?
    a. Semen
    b. Testosterone
    c. Sperm
    d. Urine

13. An IUD is:
    a. a contractible disease: inflammatory urethral disease.
    b. a means of contraception.
    c. a means of conception.
    d. gamete deformity.

14. When the loose skin covering the tip of the penis is removed, whether in infancy or adulthood, the male is then referred to as being:
    a. circumcised.
    b. impotent.
    c. aroused.
    d. neutered.

15. At what point is the developing human referred to as a fetus?
    a. At fertilization
    b. The eight-cell stage after fertilization
    c. At implantation
    d. Eight weeks after fertilization until birth

16. How many sperm can fertilize the egg?
    a. 2
    b. 1
    c. 42
    d. 23 pairs

17. The fetus floats in a fluid called:
    a. amniotic.
    b. vestibular.
    c. semen.
    d. embryonic.

18. Approximately at what age does sperm production end?
    a. 45
    b. 55
    c. 65
    d. Death

19. Where is the prostate gland located?
    a. In the glans penis
    b. In the scrotum
    c. Under the urinary bladder
    d. Lateral to the cervix

20. Which of the following layers of the uterus sheds when a woman has her period?
    a. Myometrium
    b. Perimetrium
    c. Functional layer of the mucosa
    d. Basal layer of the endometrium

21. The process of sorting chromosomes so that each gamete (egg or sperm) gets the right number of copies of the genetic material is:
    a. meiosis.
    b. fertilization.
    c. production division.
    d. mitosis.

22. A surgeon wants to treat a tumor in the myometrium by occluding the arteries that serve that layer, without affecting the endometrium, perimetrium, or other pelvic structures. She will then attempt to occlude which arteries?
    a. Arcuate arteries
    b. Common iliac arteries
    c. Straight radial arteries
    d. Spiral radial arteries

23. Which structure of the female anatomy is similar to the penis in that it becomes engorged with blood during sexual arousal?
    a. Breast
    b. Ovaries
    c. Mons pubis
    d. Clitoris

24. What kind of feedback on the hypothalamus does estrogen exert before and after ovulation, respectively?
    a. Negative/negative
    b. Negative/positive
    c. Positive/positive
    d. Positive/negative

25. Where do sperm mature?
    a. Epididymis
    b. Prostate
    c. Sertoli
    d. Penile shaft

26. An extra chromomosome 21 (trisomy 21) causes this disorder:
    a. androgen insensitivity.
    b. Down syndrome.
    c. fragile X syndrome.
    d. autism.

27. One of the most common causes of female infertility is:
    a. endometriosis.
    b. erectile dysfunction.
    c. PMS.
    d. ectopic pregnancy.

28. Symptoms like depression, anger, breast tenderness, bloating, and headache, which occur 5 days before menstruation and disappear during menstruation, are diagnostic for:
    a. PMDD.
    b. PMS.
    c. PCOS.
    d. endometriosis.

29. This disorder is related to infection with the human papilloma virus:
    a. breast cancer.
    b. prostate cancer.
    c. colon cancer.
    d. cervical cancer.

30. This condition is characterized by bleeding late in pregnancy and is often treated with a cesarean section:
    a. ectopic pregnancy.
    b. placenta previa.
    c. miscarriage.
    d. infertility.

31. This disorder is the most common cause of lower urinary tract symptoms in men:
    a. prostate cancer.
    b. benign prostatic hypertrophy.
    c. testicular cancer.
    d. androgen insensitivity syndrome.

32. Erectile dysfunction disorder is characterized by failure of:
    a. erection.
    b. ejaculation.
    c. spermatogenesis.
    d. Both a and b

33. The primary cause of male infertility is:
    a. impotence.
    b. erectile dysfunction.
    c. failure of sperm production.
    d. benign prostatic hypertrophy.

34. In androgen insensitivity syndrome, the child is:
    a. genetically female.
    b. genetically male.
    c. normal by the time they reach puberty.
    d. normal as a child but sterile as an adult.

35. Mammography is used to detect:
    a. breast cancer.
    b. cervical cancer.
    c. prostate cancer.
    d. testicular cancer.

MATCHING EXERCISES

Set 1

Please match each term with the appropriate definition.

_____	1. Cryptorchidism	a.	The placenta tears away from the uterine walls
_____	2. Mastectomy	b.	Fertilized egg implants in fallopian tubes
_____	3. Vasectomy	c.	Absence of menstruation
_____	4. Mastitis	d.	When the testes do not descend during late fetal development
_____	5. Ectopic pregnancy	e.	Severing or tying off the vas deferens; a form of birth control
_____	6. Abruptio placentae		
_____	7. Vaginitis	f.	Inflammation of the breast tissue
_____	8. Breech	g.	Inflammation of the vagina
_____	9. Dysmenorrhea	h.	Fetus coming through the birth canal buttocks first
_____	10. Amenorrhea	i.	Difficult menstruation
		j.	Removal of breast usually because of cancer

Set 2

Please match each structure with the appropriate definition.

_____	1.	Gonads
_____	2.	Areola
_____	3.	Granulosa
_____	4.	Vestibule
_____	5.	Fimbria
_____	6.	Corpus albicans
_____	7.	Sertoli
_____	8.	Tunica vaginalis
_____	9.	Inguinal
_____	10.	Tunica albuginea

a. Space between labia minora where the ure-thra and vagina empty

b. Most superficial layer of connective tissue surrounding testes

c. General term for both the ovaries and testes

d. Fibrous capsule covering the ovaries

e. Canal where the vas deferens passes from scrotum to trunk

f. Helper cells for the sperm

g. Nipple

h. Helper cells surrounding the primary oocyte

i. Ciliated projection on the distal portion of both uterine tubes

j. A degenerating structure in the ovaries

Set 3

Please match each term with the appropriate function.

_____	1.	Semen
_____	2.	Prolactin
_____	3.	Oxytocin
_____	4.	Testosterone
_____	5.	Progesterone
_____	6.	Estrogen
_____	7.	Gonadotropin-releasing hormone
_____	8.	Follicle-stimulating hormone
_____	9.	Luteinizing hormone
_____	10.	Human chorionic gonadotropin

a. Rising levels stimulate proliferation of the uterine lining

b. Rising levels maintain the buildup of the endometrium

c. Hormone responsible for maintaining the corpus luteum

d. Substance containing sperm, mucus, sugars, and certain chemicals

e. Hormone-regulating contraction of uterus and ejection of milk

f. Responsible for masculinization at puberty

g. Regulates production and secretion of cer-tain pituitary hormones

h. Initiates the development of primary follicle

i. In females, a surge in this hormone is coupled with ovulation

j. Hormone regulating the production of milk

Set 4

Please match each disorder with the appropriate treatment.

_____	1.	PMS
_____	2.	Endometriosis
_____	3.	Polycystic ovarian syndrome
_____	4.	Ectopic pregnancy
_____	5.	Breast cancer
_____	6.	Erectile dysfunction disorder
_____	7.	Benign prostatic hypertrophy
_____	8.	Prostate cancer
_____	9.	Male infertility
_____	10.	Cryptorchidism

a. Surgery to remove testicle
b. No effective medication
c. Treat symptoms, diet, anti-depressants, hormone therapy, pain relief
d. Medication, surgery
e. Oral contraceptive, hormones, pain relief, surgery
f. Medication, prosthetics, counseling
g. Termination of pregnancy, treat symptoms
h. Mastectomy, lumpectomy, chemotherapy, radiation
i. Prostatectomy, chemotherapy
j. Symptom relief, hormones, cholesterol medication, medication for insulin resistance

FILL IN THE BLANK

Fill in the blanks to complete the following statements.

1. The typical genetic makeup of humans is that females have the sex chromosomes _____ and males have _____.

2. The urethra in males transports both _____ and _____.

3. The production of sperm is termed _____.

4. A(n) _____ refers to the expansion of the penis on sexual arousal.

5. Milk-secreting sacs in the mammary lobules are called _____.

6. Between the two halves of the labia majora is an opening known as the _____ cleft.

7. The secretory phase of menstruation is also known as the _____ phase.

8. Sperm, in one of the many ducts, pass by the _____ just before flowing into the ejaculatory duct.

9. A primary oocyte has _____ chromosomes.

10. In males, luteinizing and follicle-stimulating hormones are produced and secreted by the _____.

11. If not surgically removed, the loose tissue called _____ normally covers the tip of the penis.

12. The isthmus of the uterine tubes is connected to the _____ of the uterus.

13. The primary male genitalia is/are the _____.

14. In humans, testosterone is first secreted _____.

15. The valvelike structure of the uterus called the _____ protrudes into the vagina, and its characteristic dilation marks a certain stage in delivery.

16. Cervical cancer can often be detected by a screening test known as a(n) _____.

17. In an ectopic pregnancy the embryo implants in the _____.

18. _____, a reproductive disorder, has similar symptoms to metabolic syndrome.

19. A woman with five symptoms, at least one of them affective, for most of the cycles in the previous year may be diagnosed with _____.

20. An "epidemic" of _____ in the 1980s was stopped when companies changed the material from which they made tampons.

21. Erection and ejaculation must occur together. True or false? _____

22. The underlying cause of erectile dysfunction may be physical or _____.

23. In men over 50 with symptoms of urinary frequency and an enlarged prostate, it is important to distinguish between _____ and _____.

24. Viagra is a popular medication for _____.

25. Male sterilization is accomplished by a(n) _____ for birth control purposes.

SHORT ANSWER

1. Contrast the terms *menopause* and *menarche.*

   _____

   _____

   _____

   _____

2. Describe the three stages of labor.

   _____

   _____

   _____

   _____

3. Contrast the effects estrogen and progesterone have on the endometrium.

   _____

   _____

   _____

   _____

4. What determines the sex of a baby?

   _____

   _____

   _____

   _____

5. Trace the events of ejaculation from the scrotal sac to the release of semen into the vagina.

   _____

   _____

   _____

   _____

6. Why is endometriosis so hard to treat?

   _____

   _____

   _____

   _____

7. What disorders should be ruled out in men with urinary frequency
    and urgency?

    _____

    _____

    _____

    _____

LABELING ACTIVITIES

1. Label the parts of the male reproductive system using Figure 17–10 as
    your guide.

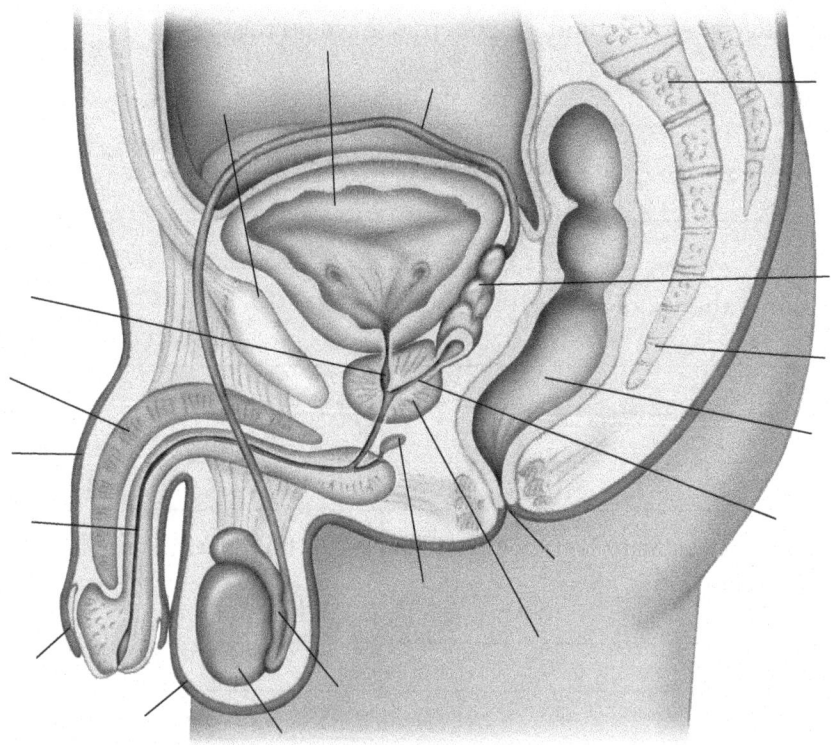

2. Label the parts of the female reproductive system using Figure 17-3 as
    your guide.

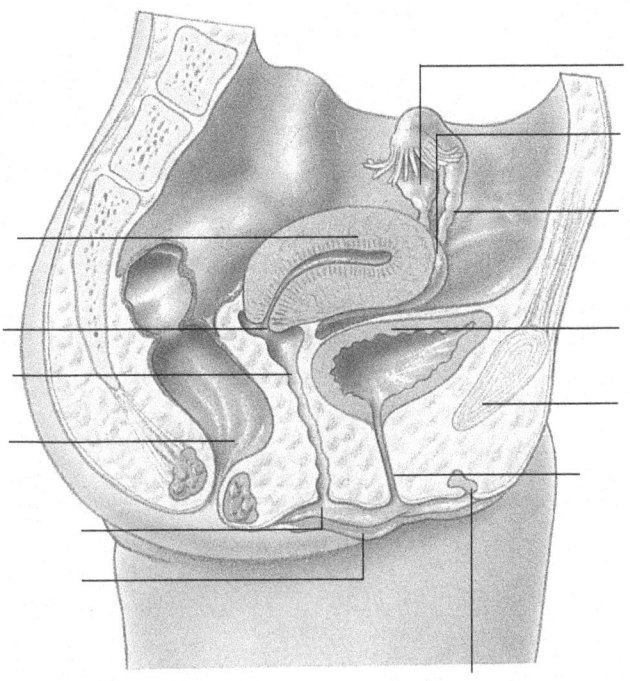

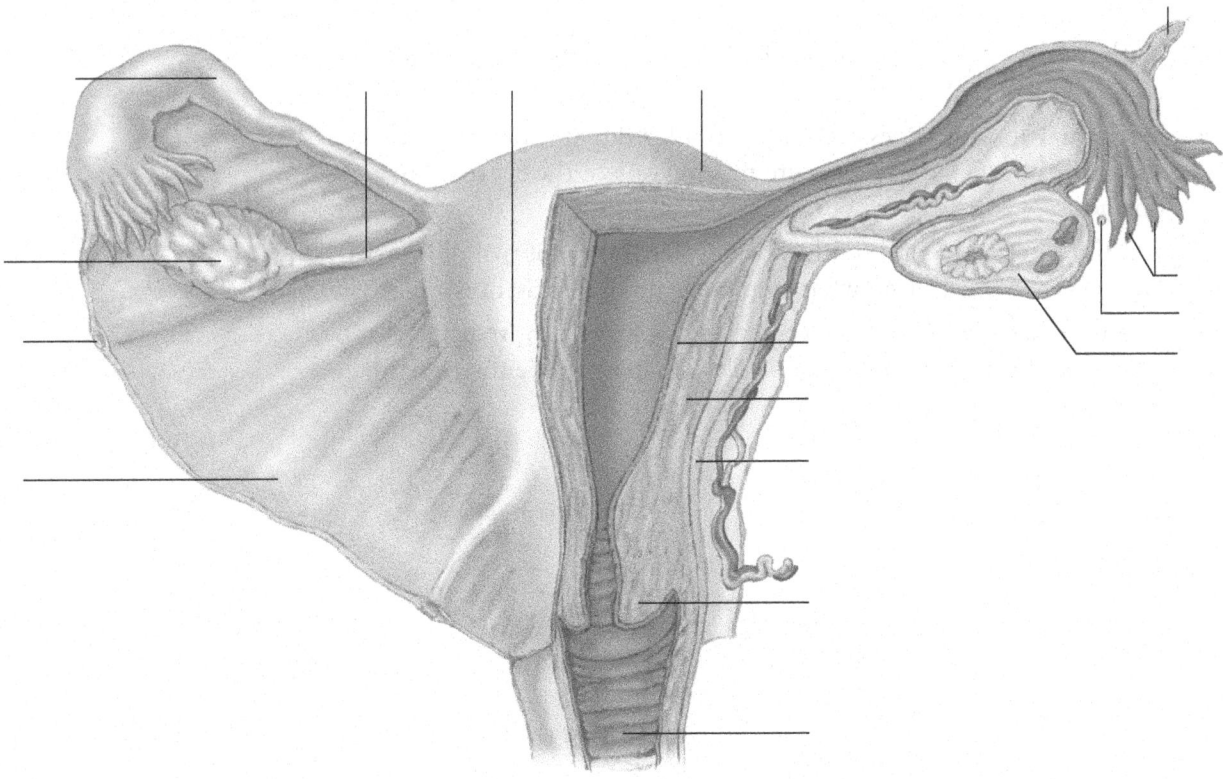

CASE STUDY

Jody and her husband, Phil, have been trying to have children for more than two years, but have been unsuccessful, so they decide to consult a fertility specialist.

1. What tests will the specialist run on Phil and Jody?

    _____

    _____

    _____

Jody's tests are all normal, but Phil has low sperm count with low motility and many abnormal sperm.

2. What are their options for fertility treatment?

    _____

    _____

    _____

LEARNING ACTIVITIES

1. List as many similarities and differences as you can between the physiology of the male and female reproductive systems. Are there any similarities?

2. Without using your book, trace the path of a sperm from testes through ejaculation.

3. Hormonal abnormalities often lead to infertility in women. Speculate about the effects if a woman were missing one of the reproductive hormones. Use the Internet to check your answer.

4. Use the Internet to find out how the typical birth control pill works. Does it make sense given what you know about female hormones?

5. You have been given the task of developing a male contraceptive pill. What hormone would you block? What would be the potential side effects?

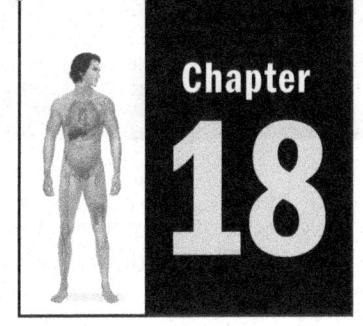

Chapter

18

BASIC DIAGNOSTIC TESTS: WHAT DO THE TESTS TELL US?

 ## CHAPTER SUMMARY

A particular sign, symptom, or syndrome can often be caused by more than one disease. To treat the patient it is important to have the correct diagnosis. Comparing the results for a patient to the normal values of a series of tests can often aid a physician in diagnosing the cause of a patient's illness, either confirming or ruling out an underlying disease. It is important to remember, however, that the diagnosis must be based on the patient and not just on test numbers.

Blood, a connective tissue, has several vital functions in your body, and because it is a transport tissue, illness will often cause some sort of change in the blood. Blood tests center around either blood cells or blood chemistry. Standard blood testing includes a complete blood count (CBC). A CBC tests the red blood count cell, percent of cells that are red blood cells (hematocrit), hemoglobin levels, white blood cell count, differential white count, and platelet count. Standard blood testing may also include other tests, such as prothrombin time and partial thromboplastin time, which measure clotting ability, and blood chemistry tests such as blood urea nitrogen (BUN), which measures the amount of waste in your blood, electrolyte levels, which measure your ion balance, and enzyme levels, which may indicate tissue damage.

Urine, filtered from you blood by your kidneys, is the by-product of an efficient fluid and waste management system. The kidneys are responsible for controlling the fluid and ion balance of your body and of removing some wastes, like urea, from your blood. Thus, the results of urine testing are often useful in diagnosing disorders. The levels of many important urine constituents can be measured using a dipstick. Dipsticks change color when exposed to urine and can be read based on a color scale. Some dipstick tests include tests for glucose, ketones, pH, and protein, all important indicators of overall health. These tests can even be performed at home. More specific, detailed follow-up tests are performed during a laboratory urinalysis. Standard urinalysis includes specific gravity (a ballpark measure of urine concentration), concentration (measured when a patient has been without water prior to the test), color (abnormal color indicates abnormal urine chemistry), odor (chemical abnormalities often cause a distinctive odor), pH (again an indicator of chemical abnormality), turbidity (cloudiness), sugar (glucose in urine indicates increased blood glucose), protein (there should be none in urine), ketone bodies (presence indicates high fat metabolism, which could indicate metabolic problems), sediment (type of sediment may aid in diagnosis), and bacteria (presence indicates infection).

Feces, the by-product of your digestive system's waste disposal, may also yield much-needed clues for diagnosis. The amount, consistency, form, shape, pH, color, and presence of blood and/or mucus in the stool all indicate something about GI tract health.

More specialized testing, depending on a patient's symptoms, may include cerebrospinal fluid testing, which tests for abnormalities in the fluid surrounding the brain and spinal cord; cardiac diagnostic tests, which test cardiac function in a variety of situations; pulmonary function tests, which examine air flow into and out of the lungs; polysomnography, which tests patients while they are sleeping; and endoscopic examination. Endoscopic examination involves the insertion of a light tube with an eyepiece into body cavities. Types of endoscopy include otoscopy (ears), laparoscopy (abdominal cavity), bronchoscopy (lungs), cystoscopy (urinary bladder), colonoscopy (large intestine), and gastroscopy (stomach).

In addition, if a patient is suspected of having a bacterial infection, often the bacteria will be tested for their antibiotic sensitivity. These tests are called culture and sensitivity tests.

CHAPTER OUTLINE

I. Blood testing
   A. Blood basics
      1. Types of cells
      2. Drawing blood
      3. Blood disorders
   B. Tests
      1. Cell numbers/function
         a. Complete blood count
            (1) Red cell count
            (2) Hematocrit
            (3) Hemoglobin
            (4) White blood cell count
            (5) Differential leukocyte count
            (6) Platelet count
         b. Prothombin time
         c. Partial thromboplastin time
      2. Blood chemistry
         a. Blood urea nitrogen
         b. Electrolytes
         c. Enzymes

II. Urine testing
   A. Catch methods
   B. Dipstick tests
   C. More specific tests
      1. Specific gravity
      2. Concentration
      3. Urine color
      4. Urine odor
      5. Urine pH
      6. Turbidity
      7. Glucose
      8. Protein
      9. Ketones
      10. Bacteria
      11. Sediment

III. Other tests
   A. Fecal testing
      1. Amount, consistency, form, shape
      2. pH
      3. Color
      4. Blood
      5. Mucus
   B. Cerebrospinal fluid testing
      1. Color
      2. Cell counts
   C. Culture and sensitivity testing
   D. Cardiac diagnostics
      1. Electrocardiogram
      2. Holter monitor
      3. Stress testing
   E. Endoscopy
      1. Otoscopy
      2. Bronchoscopy
      3. Gastroscopy
      4. Laparoscopy
      5. Cystoscopy
      6. Colonoscopy
   F. Pulmonary function tests
   G. Polysomnography

MEDICAL TERMINOLOGY REVIEW

Define the following terms.

1. Anemia: _____

2. Hematocrit: _____

3. Hemoglobin: _____

4. Leukocytosis: _____

5. Leukopenia: _____

6. Thrombocytopenia: _____

7. Electrocardiogram: _____

8. Holter monitor: _____

9. Endoscope: _____

10. Laparoscope: _____

MULTIPLE CHOICE

Circle the letter of the correct answer.

1. Red blood cells are also known as:
   a. erythrocytes.
   b. leukocytes.
   c. thrombocytes.
   d. platelets.

2. A condition of too few red bloods cells is known as:
   a. leukemia.
   b. thrombocytopenia.
   c. anemia.
   d. polycythemia.

3. A complete blood count usually includes the following:
   a. platelet count.
   b. hematocrit.
   c. red blood cell count.
   d. All of the above

4. Which blood cells is the most numerous in a blood sample?
   a. Erythrocytes
   b. Leukocytes
   c. White blood cells
   d. Platelets

5. Normal hemoglobin values for women are:
   a. 20-25 g/dL.
   b. 12-16 g/dL.
   c. 60-70 g/dL.
   d. 45-50 g/dL.

6. The most common white blood cells are:
   a. lymphocytes.
   b. basophils.
   c. monocytes.
   d. neutrophils.

7. An abnormally high BUN is often a sign of:
   a. bone fracture.
   b. high carbohydrate diet.
   c. liver failure.
   d. kidney disease.

8. Which of the following is *not* considered an electrolyte?
   a. Glucose
   b. Sodium
   c. Calcium
   d. Chloride

9. Why might a test of blood enzyme levels be useful in diagnosing an illness?
   a. Enzymes are proteins.
   b. Enzymes may be released due to cell injury or death.
   c. Low enzyme levels are diagnostic for viral infection.
   d. Why not? You might as well have as much information as possible.

10. In the urine, glucose levels should be approximately:
   a. 100 mg/dL.
   b. 0 mg/dL.
   c. 15 mg/dL.
   d. It depends on the age of the patient.

11. Specific gravity testing looks at the ability of the kidney to _____.
   a. make urine
   b. control urine pH
   c. concentrate urine
   d. hydrate the urine

12. Ketones should not typically be present in urine. Their presence indicates:
   a. kidney stones.
   b. increased fat metabolism.
   c. emotional stress.
   d. exercise.

13. Protein in urine often indicates damage to which organ(s)?
   a. Brain
   b. Liver
   c. Pancreas
   d. Kidneys

14. Watery fecal matter is called:
   a. constipation.
   b. celiac disease.
   c. diarrhea.
   d. hydrostool.

15. Normal stool color is:
   a. brown.
   b. yellow.
   c. green.
   d. black.

16. Gastric bleeding will often turn a stool this color:
   a. red.
   b. black.
   c. yellow.
   d. green.

17. CSF is normally:
   a. cloudy.
   b. yellow.
   c. colorless.
   d. red.

18. Significant numbers of white blood cells in CSF typically indicate:
   a. inflammation or injury to the nervous system.
   b. anemia.
   c. brain tumor.
   d. nothing; it's normal.

19. The purpose of culture and sensitivity testing is:
   a. testing drug sensitivity of pathogens.
   b. checking hormone levels.
   c. preventing biological terrorism.
   d. traveling to foreign countries.

20. Long-term monitoring of heart activity is done using a(n):
   a. EKG.
   b. stress test.
   c. cardioscope.
   d. Holter monitor.

21. This type of test involves walking on a treadmill:
   a. stress test.
   b. EKG.
   c. PFT.
   d. Holter monitor.

22. A tube that examines the lungs is a(n):
   a. otoscope.
   b. endoscope.
   c. bronchoscope.
   d. cystoscope.

23. _____ is recommended for people over 50 or younger patients with a family history of colon cancer.
   a. Bronchoscopy
   b. Cystoscopy
   c. Colonoscopy
   d. Otoscopy

24. If you are suspected of having sleep apnea, you may be sent for this type of testing:
    a. colonoscopy.
    b. polysomnography.
    c. urinalysis.
    d. culture and sensitivity.

25. _____ looks at flow and volume of air into and out of the lungs.
    a. PFT
    b. Bronchoscopy
    c. Laryngoscopy
    d. Pleuroscopy

MATCHING EXERCISES

Set 1

Please match each test with what it measures.

_____ 1. Hematocrit
_____ 2. Hemoglobin
_____ 3. Differential
_____ 4. CBC
_____ 5. Pro time
_____ 6. BUN
_____ 7. Electrolytes
_____ 8. Enzymes
_____ 9. Glucose
_____ 10. Platelet count

a. Blood oxygen transport molecule in blood
b. Waste product in blood
c. Indication of tissue damage
d. RBCs in specific volume of blood
e. Levels of prothrombin
f. Blood sugar
g. Number of specific white blood cells
h. Ions in blood
i. RBC, Hct, Hgb, WBC, Diff, Platelets
j. Platelets in specific volume of blood

Set 2

Please match each test with what it measures.

_____ 1. Specific gravity
_____ 2. pH
_____ 3. Turbidity
_____ 4. Glucose
_____ 5. Ketones
_____ 6. Bacteria
_____ 7. Sediment
_____ 8. Stool color
_____ 9. Stool consistency
_____ 10. CSF testing

a. Indicates fat metabolism
b. Indicates infection
c. Indicates hydration level
d. Solids in urine
e. Indicative of diet or digestive tract health
f. Acidity
g. Sugar
h. Test fluid around nervous system
i. Cloudiness
j. Usually dependent on diet

Set 3

Please match each abnormality with its cause.

_____	1.	Increased red blood cells	a.	Diuretic side effect
_____	2.	Low hematocrit	b.	Diabetes mellitus
_____	3.	Decreased white blood cells	c.	Gastric ulcer
_____	4.	Increased BUN	d.	Overhydration
_____	5.	Decreased potassium	e.	Chronic infection
_____	6.	Sweet smelling urine	f.	Urinary tract infection
_____	7.	Decreased urine concentration	g.	Brain infection
_____	8.	Bacteria in urine	h.	Hemorrhage
_____	9.	Blood in stool	i.	Kidney disease
_____	10.	White blood cells in CSF	j.	Dehydration

FILL IN THE BLANK

Fill in the blanks to complete the following statements.

1. A diagnosis is based solely on the test numbers. True or false?

   _____

2. The _____ are the blood
   vessels normally used for blood samples.

3. The test used to determine the number of red blood cells in a specific
   volume of blood is the

   _____.

4. List one of the possible causes of anemia:

   _____

5. This molecule transports oxygen around the body:

   _____.

6. An increase in this white blood cell may indicate an allergic reaction.

   _____

7. _____ is the name for any
   condition in which total white blood cells count falls below $4,000/mm^3$.

8. The molecule _____ is
   formed by the liver as waste product of protein catabolism.

9. Standard urine testing is often done using this measurement tool.

   _____

10. Urine should typically be this color:

    _____

11. Cloudiness of urine is known as

    _____.

12. Urine should never have

    _____ in it.

13. Electrical measurements of the heart are made using a(n)

_____.

14. A(n) _____ is the general
term for a device used to look inside the body.

15. Gallbladder removal is now often being performed in a minimally
invasive way using a(n)

_____.

SHORT ANSWER

1. List the types of blood cells and their functions.

_____

_____

_____

_____

2. What tests would you perform to diagnose a patient's blood clotting
abnormality?

_____

_____

_____

_____

3. A patient has a fainting spell and is transported to the emergency department. A blood test records elevated enzymes. What other tests would you
order?

_____

_____

_____

_____

4. List the scopes and the parts of the body they examine.

_____

_____

_____

_____

5. What tests are part of the CBC? List and briefly describe them.

_____

_____

_____

_____

CASE STUDY

An elderly woman presents at the emergency department having collapsed at a birthday party. When she regains consciousness, she is confused and sleepy. She cannot really respond to questions. She has an obvious bump on her head above her left ear, but other partygoers say she was "not herself" even before her collapse. They report she has not been drinking alcohol and to their knowledge does not have any chronic illness. When questioned about what she might have ingested at the party, they confess that they do not remember her eating or drinking anything all day. She is obviously dehydrated and is running a fever.

1. What test(s) should you run initially?

    _____

    _____

Her CBC shows elevated RBCs, WBCs, and Hct. The differential count shows elevated neutrophils, basophils, monocytes, and eosinophils. Her blood chemistry shows elevated BUN, calcium, chloride, potassium, and sodium, but normal enzymes. X-rays and MRI scan of her head show no skull fracture or brain injury.

2. What is your initial diagnosis?

    _____

    _____

3. Given these results, what test(s) should you run next?

    _____

    _____

Urinalysis shows bacteria, blood, and protein in the urine. Her urine is extremely concentrated, brown, cloudy, has an unpleasant odor and significant ketone bodies. Fecal testing is normal (if you ran it). CSF is normal (if you ran it).

4. What is your final diagnosis?

    _____

    _____

5. How would you treat the patient?

    _____

    _____

LEARNING ACTIVITIES

1. Use some urine dipsticks to test the effect of what you eat on urine composition. After a meal, test your urine following the directions on the package. Then test it first thing in the morning after you have been asleep and haven't had anything to drink in a while. Is it different? How about after you drink a caffeinated soda? Or after you eat a lot of sweet snacks? Can you see any difference? (One warning: keep in mind that these tests are not done under clinical conditions! They do not necessarily say anything about your health. Do not ever experiment on yourself without permission of your doctor and your parents.)

2. Learn the abnormal test values for particular diseases. Either use the ones in the chapter, or do some research. Discuss your findings.

3. The next time you go to the doctor, ask about the tests she orders. What do they tell your physician about your health?

4. There are many other tests that we have not described here. For one of the disorders in the chapter, research what other tests might be ordered to aid diagnosis.

5. Many tests can be influenced by diet, medications, or physical activity. For one test, research the influences that might alter the results.

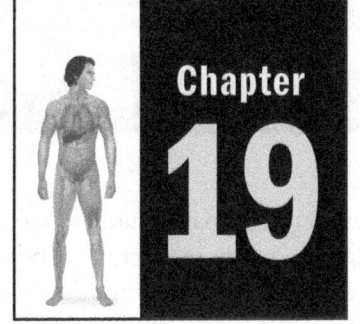

ANATOMY AND PHYSIOLOGY AND THE SCIENTIFIC METHOD

 ## CHAPTER SUMMARY

The scientific method is a way of gathering empirical knowledge, facts, and observations of natural and physical phenomena. It consists of identifying research questions, formulating hypotheses, and testing them. Scientific hypotheses must rely on natural or physical explanation and must be refutable given sufficient evidence. Hypotheses are never proven true, only rejected or supported. If a hypothesis is repeatedly tested and stands up to scrutiny, it may become incorporated into larger explanations called theories. Scientific theories are the result of rigorous experimentation and investigation by multiple, independent researchers. They are very well supported by existing data, but may one day be disproven in light of new data.

Hypotheses can be tested directly or indirectly, when the hypotheses explain phenomena that are difficult to observe or happened in the past. Indirect testing is done by generating predictions and testing those predictions instead. For example, some bacteria are difficult to screen for in blood samples, but their presence can be inferred if white blood cell counts are high.

There are generally three types of investigations used to test hypotheses: descriptive, comparative, or experimental. Descriptive hypotheses are conducted by making observations or collecting data. While a hypothesis may not be explicitly tested with this form of investigation, the descriptions they offer can be evaluated for their consistency with the hypotheses. Comparative investigations are used to test hypotheses by making comparisons between groups, populations, or species that differ. Experiments involve manipulating one or more independent variables, using proper controls, and looking for changes in dependent variables.

Collecting data to test hypotheses requires careful consideration of the methodologies used, necessary controls, and safety concerns. While both qualitative and quantitative measurements can be used to test hypotheses, the best quantitative measurements are both accurate and precise.

A data set can be quantitatively summarized using one or more values that describe the typical value in the data set: the mode, median, or mean. Data can also be visually summarized using a variety of figures and tables including bar graphs, scatter plots, and summary tables. Knowing what type of figure or table to use depends on the specific data you wish to present, whether the variables are measured continuously or categorically, and whether the data come from a single subject or multiple subjects.

Drawing inferences can be done by comparing samples visually or using statistical tests. Statistical tests provide a more objective means of determining whether or not two samples are different, but they can never definitively prove that there is a difference. There is always a chance of a false positive. Often, researchers produce models, simple mathematical equations or figures that summarize a data set, to be able to make predictions about values outside the data set. Using models must be done carefully as models have important limitations.

Scientists typically communicate results to one another using scientific journals, books, or speeches at scientific conferences. These results are then summarized by scientists or science journalists for the general public, potentially introducing errors, oversimplifying the results, or misrepresenting what the study actually showed.

Although science has been crucial in shaping our modern world, it is not the only pattern of knowledge and it has its limitations, chiefly that scientific hypotheses can never be definitively proven true and may change over time in light of new data.

CHAPTER OUTLINE

I. What is the scientific method?
 A. Four patterns of knowing
  1. Empirical
  2. Personal
  3. Aesthetics
  4. Ethics
 B. The scientific method
  1. Formulating questions
  2. Hypotheses and theories
  3. Methods
  4. Results
  5. Conclusions

II. Testing hypotheses
 A. Types of investigations
  1. Descriptive
  2. Comparative
  3. Experimental
 B. Important considerations
  1. Selecting equipment and technology
  2. Safety and disposal
 C. Making measurements
  1. Quantitative and qualitative data
  2. Accuracy and precision
  3. Mode, median, and mean
 D. Presenting Data
  1. Bar graphs
  2. Scatter plots
  3. Summary tables
  4. Captions
  5. Common graphing errors

      E. Drawing inferences
          1. Statistical tests
      F. Evaluating and using models
      G. Communicating scientific results
  III. Limitations of the scientific method

TERMINOLOGY REVIEW

Define the following terms.

1. Empirical knowledge: _____

2. Hypothesis: _____

3. Indirect testing: _____

4. Independent variable: _____

5. Biohazard: _____

6. Quantitative data: _____

7. Precision: _____

8. Median: _____

9. Caption: _____

10. Model: _____

MULTIPLE CHOICE

Circle the letter of the correct answer.

1. Scientific studies yield _____ knowledge.
   a. Empirical
   b. Personal
   c. Aesthetic
   d. Ethics

2. Religious or moral beliefs factor in one's _____ knowledge.
   a. Empirical
   b. Personal
   c. Aesthetic
   d. Ethics

3. A _____ is an educated guess used to explain a phenomenon or answer a question.
   a. Question
   b. Hypothesis
   c. Method
   d. Inference

4. _____ testing involves making measurements and observations of the test subjects, experiment, or patient.
   a. Direct
   b. Indirect
   c. Inferential
   d. Categorical

5. _____ investigations require the use of carefully considered and designed control groups.
   a. Descriptive
   b. Comparative
   c. Experimental
   d. All of the above.

6. Which of the following would be considered a physical hazard?
   a. A machine with moving parts
   b. An unused scalpel
   c. A wet floor
   d. All of the above.

7. Which of the following would be considered a chemical hazard?
   a. An open flame
   b. A bottle of hydrochloric acid
   c. A used bandage
   d. None of the above.

8. A bloody bed sheet in a hospital should be considered a _____.
   a. Physical hazard
   b. Chemical hazard
   c. Biohazard
   d. None of the above.

9. Which of the following would be an example of qualitative data?
   a. Height in inches
   b. Weight in pounds
   c. Taste (salty, bitter, or sweet)
   d. Blood pressure (mmHg)

10. A device has high _____ if it is likely to provide the same measurement repeatedly.
    a. Accuracy
    b. Precision
    c. Specificity
    d. Mode

11. The mode of the sample {4, 8, 2, 6, 2} is _____.
    a. 2
    b. 4
    c. 4.4
    d. 8

12. The median of the sample {4, 8, 2, 6, 2} is _____.
    a. 2
    b. 4
    c. 4.4
    d. 8

13. The mean of the sample {4, 8, 2, 6, 2} is _____.
    a. 2
    b. 4
    c. 4.4
    d. 8

14. Which of the following graphs is used to display the means of different samples for comparison?
    a. Bar
    b. Scatter plot with connected dots
    c. Scatter plot with best fit lines
    d. None of the above.

15. Which of the following graphs is used to display the average relationship between two continuous variables?
    a. Bar
    b. Scatter plot with connected dots
    c. Scatter plot with a best fit line
    d. None of the above.

16. Which of the following graphs is used to display data from a single participant or subject when both variables are continuous?
    a. Bar
    b. Scatter plots with connected dots
    c. Scatter plots with best fit lines
    d. None of the above.

17. Which of the following should be included with a graph?
    a. Axis labels
    b. Units
    c. A legend when appropriate
    d. All of the above.

18. Which of the following should be considered when using models?
    a. Models only describe an average relationship.
    b. Models are often based on a sample of data that may be small in comparison to the population the model describes.
    c. Models are typically only useful in making predictions within the range of values seen in the data set.
    d. All of the above.

19. Scientists typically communicate the results of their research to other scientists via:
    a. Social media
    b. Scientific journals
    c. Newspaper articles
    d. Newsletters

20. Which of the following could be considered limitations of the scientific method?
    a. Scientific explanations can never be proven definitively true.
    b. Scientific explanations will change over time in light of new tests and data.
    c. Explanations that require invoking supernatural forces or breaking laws of nature cannot be tested or considered using the scientific method.
    d. All of the above.

 ## MATCHING EXERCISES

Set 1
Please match each term with the appropriate example.

_____ 1. Empirical knowledge

_____ 2. Personal knowledge

_____ 3. Aesthetics

_____ 4. Ethics

a. A patient refuses a blood transfusion based on their religious beliefs

b. A nurse gives a glass of water because she remembers patients that took this drug before got very thirsty afterwards

c. A doctor diagnoses a disease by ruling out alternatives and testing predictions

d. A doctor carefully considers how to discuss a prognosis with a patient by considering how they would feel in the patient's position

Set 2
Please match each term with the appropriate example.

_____ 1. Descriptive investigations

_____ 2. Comparative investigations

_____ 3. Experimental investigations

a. Comparing children at different schools to see if test scores are correlated with the number of minutes of recess children receive

b. Growing plants with or without nitrogen addition to see how plant growth is affected

c. Measuring the average height of all the students in your class to describe the mean and range

Set 3

Please match each term with the appropriate example.

_____ 1. Physical hazards   a. An unused needle

_____ 2. Chemical hazards   b. A vial of blood that is leaking

_____ 3. Biohazards   c. Bleach and other cleaning products

Set 4

Please match each term with the appropriate situation.

_____ 1. Bar Graph

_____ 2. Scatter Plot

_____ 3. Line Graph

a. Used when the dependent variable is continuous but the independent is categorical

b. Used when both variables are continuous and the data points are unrelated

c. Used when both variables are continuous and the data points come from the same individual or subject

FILL IN THE BLANK

Fill in the blanks to complete the following statements.

1. Using the scientific method produces _____ knowledge under Carper's patterns of knowing.

2. The scientific method is conducted by testing one or more _____, educated guesses based on previous studies.

3. _____ is used when a hypothesis is based on phenomena that occurred in the past or are difficult to observe.

4. _____ investigations are used to make observations without necessarily testing a specific hypothesis.

5. In an experimental investigation, the _____ variable is the variable that the researchers directly manipulate to study its effect(s).

6. To rule out extraneous, confounding effects, experiments require carefully considered and designed _____.

7. _____ hazards include sources of mechanical injury, such as unused sharp instruments.

8. Soiled sheets my represent a _____ because they may contain infectious pathogens.

9. Researchers should wear appropriate _____, such as gloves, gowns, and face masks, to avoid many common hazards.

10. _____ data cannot be measured on a numerical scale, such as color or taste.

11. An instrument is considered _____ if it provides measurements that are very close to the actual value.

12. The value in a data set that is repeated the most is its _____.

13. The _____ of a data set is calculated by adding all the values and dividing by the number of the values in the data set.

14. _____ is a way of representing the amount of variation in a data set and is often used as error bars on a bar graph.

15. A _____ graph is used when the independent variable is categorical and the dependent is continuous.

16. In a scatter plot, both variables must be _____.

17. A _____ is used in a scatter plot to represent the average relationship between variables.

18. A study concludes that a drug does not raise blood pressure as a side effect when it in fact does. This is an example of a _____.

19. A _____ is a simple mathematical equation or figure constructed to describe trends in a data set.

20. A scientific _____ is a monthly or quarterly publication in a specific area of research.

SHORT ANSWER

1. What distinguishes scientific hypotheses from other types of explanations or guesses?

_____
_____
_____
_____

2. Under what circumstances would indirect testing be necessary?

_____
_____
_____
_____

3. What is the purpose of a control group in an experiment?

_____

_____

_____

_____

4. What factors should be considered when choosing equipment or technology for research?

_____

_____

_____

_____

5. Briefly list some ways researchers and health professionals can minimize risks from hazards.

_____

_____

_____

_____

6. What distinguishes a false positive from a false negative?

_____

_____

_____

_____

7. Briefly discuss two limitations of the scientific method.

_____

_____

_____

_____

GRAPHING EXERCISE

The data set below comes from repeatedly measuring the heart rate of a student in gym class as they begin to sprint on a treadmill.

Time (minutes)	Heart Rate (bpm)
0	60
2	85
4	100
6	110
8	85
10	75

1. Based on the data set above and the description, what type of graph should be used? Why?

   _____

   _____

   _____

   _____

2. Using a ruler, draw your graph on the plot below. Make sure to label your axes and include units.

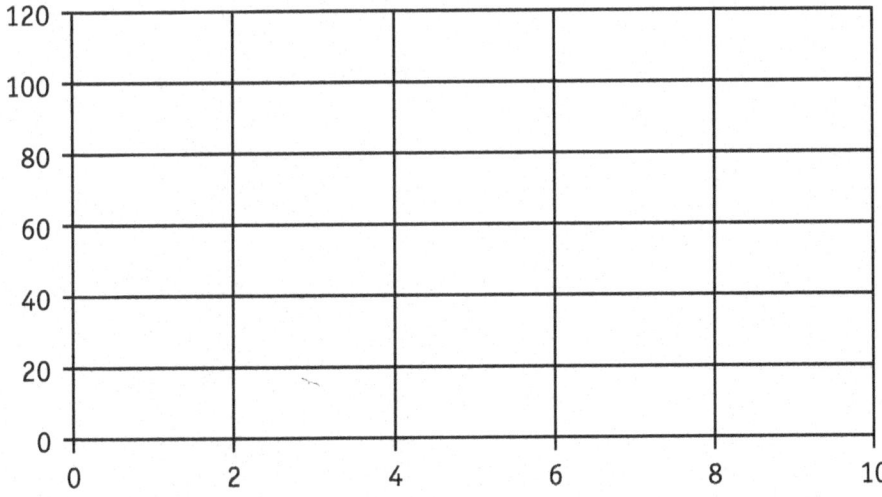

3. Write a descriptive caption for this figure (label it Figure 1). Where should this caption be placed?

   _____

   _____

   _____

   _____

LEARNING ACTIVITIES

1. Identify potential hazards in your home and come up with a list of ways to minimize them. Classify them as physical, chemical, or biohazards.

2. Find leaves in your yard, at school, or at a park and measure their lengths. Calculate the mean, median, and mode lengths.

3. Find examples of graphs in scientific journals or magazines like *Scientific American*. Identify the variables presented in the graph and critique the choice of graph and the caption. Share these critiques with your classmates.

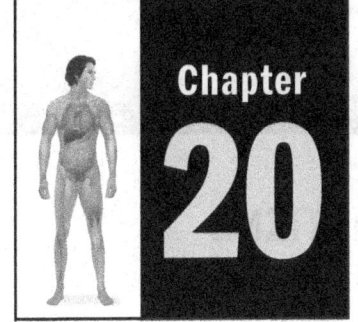

Chapter

20

THE JOURNEY'S END: NOW WHAT?

 ## CHAPTER SUMMARY

Nearly every day there is a news story featuring some aspect of anatomy, physiology, or pathology. Forensic science applies principles of anatomy, physiology, and pathology in investigations. Forensic science has been used to detect disease in ancient skeletons, to identify skeletal remains, and to convict or exonerate suspects. The tools of forensic scientists include fingerprints, skeletal anatomy, and DNA fingerprinting.

For many clinicians, their client base is aging, as is much of America. With the aging of the population comes a number of characteristics that apply to aging patients that may not be apparent in younger patients. Older patients have lost lean body mass and bone density and gained body fat. Their senses have begun to deteriorate, and there are often obvious behavioral changes. Many geriatric patients are not well medicated for pain. The cardiovascular system works less efficiently as valves harden and blood vessels harden, often leading to increased blood pressure. Cardiac output decreases. Between the ages of 20 and 80, patients lose 50% of renal function. Skin becomes more delicate. Most importantly, many geriatric patients are on several medications, often prescribed by different doctors. Drug interactions are often a risk.

Your health as you age may be influenced by lifestyle choices you make as a younger person. Lifestyle choices that minimize stress include regular physical activity and good nutrition and minimize risk factors, such as excess sun exposure. Tobacco use should be avoided and alcohol should be consumed only in moderation. Medical decisions should be made with the appropriate information at hand about treatment options and outcomes. Unprotected sex should be avoided to decrease the risk of sexually transmitted diseases.

One of the major public health issues in the United States is cancer treatment and prevention. Cancer is often caused by an interaction between genetics and environmental triggers. Everybody who smokes does not develop lung cancer, but there is no way to tell if you will trigger a potentially fatal cancer by smoking as a college student. Cancer treatment centers on destroying the cancerous cells and preventing their return. Surgery cuts out the cancer. Chemotherapy uses drugs to target rapidly dividing cells. Radiation uses energy waves to shrink tumors, and biological therapies use the body's own defenses to attack the cancer cells. Many cancers can be successfully treated if they are caught early before they have a chance to spread.

CHAPTER OUTLINE

I. Forensic science
   A. Disease detection
   B. Primitive surgery
   C. Remains identification
   D. Murder investigations
      1. Fingerprints
      2. DNA fingerprinting

II. Geriatrics
   A. General body changes
   B. Gustatory changes
   C. Nervous system changes
   D. Cardiovascular changes
   E. Urinary system changes
   F. Integumentary system changes
   G. Polypharmacy

III. Wellness
   A. Nervous system
   B. Skeletal system
   C. Muscular system
   D. Integumentary system
   E. Cardiovascular system
   F. Respiratory system
   G. Gastrointestinal system
   H. Endocrine system
   I. Sensory system
   J. Immune system
   K. Reproductive system
      1. Diet
      2. STDs

IV. Cancer prevention and treatment
   A. Causes
   B. Treatments
   C. Melanoma

V. Amazing body facts

VI. New research

MEDICAL TERMINOLGY REVIEW

Define the following terms.

1. DNA fingerprinting: _____

2. Geriatric: _____

3. Polypharmacy: _____

4. Chemotherapy: _____

5. Radiation: _____

6. Biological therapy: _____

7. Sexually transmitted disease: _____

8. Melanoma: _____

9. Colonoscopy: _____

10. Mammography: _____

MULTIPLE CHOICE

Circle the letter of the correct answer.

1. Bone scouring is an indication of:
   a. tuberculosis.
   b. hepatitis.
   c. calcium deficiency.
   d. high-impact mechanical stress.

2. Which nerve is damaged in carpal tunnel syndrome?
   a. Median
   b. Acoustic
   c. Sciatic
   d. Radial

3. What bone fragment, showing evidence of primitive surgery, was found in a 400-year-old trash dump?
   a. Skull
   b. Femur
   c. Rib
   d. Sacrum

4. Who was Josef Mengele?
   a. The Merciless Hooded Ghost responsible for the brutality against Native Americans between 1812 and 1831
   b. The Fearless Ace responsible for "downing" several Allied Force fighter jets
   c. The Consignor of Hotel Hanoi responsible for the torture of many prisoners of war during the Vietnam War
   d. The Angel of Death responsible for the deaths of many humans during World War II

5. According to your text, what substance containing thallium was intentionally and secretly fed to the victim, causing death?
   a. Rat poison
   b. Drain cleaner
   c. Paint thinner
   d. Motor oil

6. Which of the following is true about fingerprints?
   a. Only identical twins have the same fingerprints.
   b. They are friction ridges on the hands and feet.
   c. They are only fully formed between the first and second year of life.
   d. All of the above

7. DNA fingerprinting helped determine that Thomas Jefferson, the third president of the United States, or a relative:
   a. murdered his first wife.
   b. burglarized the White House repeatedly.
   c. fathered children of his slave.
   d. had Alzheimer's disease.

8. In the elderly, what changes are seen in the taste buds?
   a. The acuity of salt becomes more sensitive.
   b. The number of taste buds increases.
   c. Sweet tastes become less discernable than bitter tastes.
   d. All of the above

9. Although bone loss occurs in both men and women, the highest percentage of bone loss in women can be seen:
   a. 10 years after menopause.
   b. in the first 5 years postmenopause.
   c. 6 months prior to menopause.
   d. after age 70.

10. As we age:
    a. heart valves become soft.
    b. cardiac output decreases.
    c. blood pressure decreases.
    d. All of the above

11. What is bone scouring?
    a. The thinning of the bone's cortex as a result of poor nutrition
    b. The thickening and extension of the trabeculae as a result of weight-bearing activities
    c. The malrepair of bones as a result of improper splinting or casting of a fractured bone
    d. The destruction of smooth bone surfaces as a result of colonization of bacteria

12. In forensic science, which anatomical structure is examined for evidence of tuberculosis?
    a. The skin
    b. The nasal apertures
    c. The retina of the left eye
    d. Ends of long bones

13. In the elderly, changes in the integumentary system include:
    a. increased skin delicacy.
    b. loss of elasticity.
    c. multiple lesions.
    d. All of the above

14. Stress:
    a. is not a natural part of life and must be controlled.
    b. gives a false sense of security and protection.
    c. is good and necessary, but chronic stress may be problematic.
    d. All of the above

15. Which of the following is recommended for maintaining healthy bones?
    a. Calcium and vitamins
    b. Weight-bearing exercises
    c. Repetitive motion such as typing
    d. Both a and b

16. What may cause carpal tunnel syndrome?
    a. Bright lights and abuse of such drugs as ecstasy
    b. Weight-bearing exercises
    c. Loud sounds and high pitches
    d. Repetitive motion such as playing the piano

17. For optimum health, how much water is recommended per day?
    a. 3 glasses
    b. 8 glasses
    c. 14 glasses
    d. 1 cup

18. What technique was initially used to confirm that Wolfgang Gerhard was really Josef Mengele?
    a. Video skull–face superimposition
    b. Fingerprints
    c. Voice recognition
    d. Reverse psychology and multiple interviews

19. Which of the following has/have an adverse effect on the cardiovascular system?
    a. Alcohol
    b. Smoking
    c. Saturated fats
    d. All of the above

20. Why can high doses of vitamins A, D, E, and K actually harm the body?
    a. Being water soluble, they tax the kidneys.
    b. They deteriorate the stomach's lining.
    c. They can build up to a toxic level.
    d. They prevent absorption of proteins and carbohydrates.

21. What is needed to identify a criminal suspect through DNA fingerprinting?
    a. Both blood and reproductive fluid from a crime scene
    b. Both DNA from a crime scene and a known DNA sample
    c. Both DNA from the mother and father of the accused
    d. Only the DNA from the crime scene

22. The simplest and most effective way to stop the spread of infections is to:
    a. avoid contact.
    b. wash hands.
    c. drink more water.
    d. take antibiotics.

23. An antibiotic cream or tablet is what kind of agent?
    a. Antifungal
    b. Antiviral
    c. Antiprotozoan
    d. Antibacterial

24. Which of the following is/are considered diuretics?
    a. Water
    b. Caffeinated coffee
    c. Gatorade
    d. Both a and c

25. The ability to roll your tongue is:
    a. gender-specific.
    b. learned.
    c. an inherited trait.
    d. race-specific.

MATCHING EXERCISES

Set 1

Please match each vitamin with the appropriate function.

_____ 1. Vitamin A
_____ 2. Calcium
_____ 3. Vitamin B_1
_____ 4. Vitamin B_{12}
_____ 5. Vitamin C
_____ 6. Niacin
_____ 7. Vitamin E
_____ 8. Vitamin K
_____ 9. Vitamin D
_____ 10. Folic acid

a. Aids in the absorption of calcium from the gut
b. Facilitates fat synthesis and glycolysis
c. Recommended for proper night vision
d. Raw material for bones and teeth
e. Needed for hemolytic resistance of RBCs
f. Needed to prevent spina bifida
g. Needed for carbohydrate metabolism and normal digestion and appetite
h. To treat pernicious anemia
i. Needed for proper blood clotting
j. Aids in the absorption of iron

Set 2

Please match each disorder with the appropriate description.

_____ 1.	SIDS	a. Common among coalminers
_____ 2.	Black lung	b. Human papilloma virus
_____ 3.	Lead poisoning	c. Presents with fluid-filled vesicles on the genitalia
_____ 4.	Genital warts	d. Presents with swollen testes and inflamed cervix
_____ 5.	Syphilis	e. Skin cancer
_____ 6.	Herpes	f. Normally thought of as a bacterial pulmonary disease
_____ 7.	Gonorrhea	g. Associated with infants
_____ 8.	Chlamydia	h. From pewter plates
_____ 9.	Melanoma	i. Presents with degeneration of the nervous system
_____ 10.	Tuberculosis	j. Presents with purulent discharge and abnormal menstruation

Set 3

Please match each system with the appropriate disorder.

_____ 1.	Reproductive system	a. Incontinence
_____ 2.	Immune system	b. Pain and stress
_____ 3.	Sensory system	c. Clogged vessels
_____ 4.	Endocrine system	d. Antibiotics abuse
_____ 5.	Brain and nervous system	e. Smoking and emphysema
_____ 6.	Respiratory system	f. High level of noise
_____ 7.	Skeletal system	g. Steroid abuse
_____ 8.	Integumentary system	h. Skin cancer
_____ 9.	Cardiovascular system	i. Carpal tunnel syndrome
_____ 10.	Urinary system	j. Smoking and SIDS

FILL IN THE BLANK

Fill in the blanks to complete the following statements.

1. Ancient _____ were stricken with tuberculosis, as investigators discovered by examining their bones.

2. According to the text, a bone fragment indicative of primitive surgery was found in a 400-year-old trash dump in

    _____.

3. Administering many drugs at the same time is termed

    _____.

4. In the absence of disease, the brain continues to mature up to the age of

    _____.

5. Bone density usually reaches its greatest peak at

    _____.

6. In general, as an individual ages, he or she loses

    _____ and
    _____ and gains
    _____.

7. A pap smear is a test for cancer of the

    _____.

8. Anabolic steroids are closely related to the hormone

    _____.

9. A mammogram is a test for cancer of the

    _____.

10. Approximately_____people die annually in the United States because of smoking-related disease.

11. The cancer treatment that uses energy waves rather than chemicals to shrink tumors is called _____
    therapy.

12. Clients who have the procedure called

    _____ done usually were 26 percent less likely to have their cancer return within 5 years than clients who only had the tumor removed.

13. Smoking can lead to chronic respiratory diseases such as

    _____,
    _____, and
    _____.

14. CIPA is the acronym for

    _____.

15. According to your text, thallium was found in the victim's
    _____, which linked the wife to the murder.

SHORT ANSWER

1. How does the sexually transmitted disease HPV present itself in both males and females?

   _____

   _____

   _____

   _____

2. Besides completely staying out of the sun, in what ways can we prevent excessive sun exposure?

   _____

   _____

   _____

   _____

3. What are the serious side effects of steroid abuse in both men and women?

   _____

   _____

   _____

   _____

4. In what ways can untreated pain affect the elderly?

   _____

   _____

   _____

   _____

5. Why, during the Middle Ages, were tomatoes believed to be poisonous?

   _____

   _____

   _____

   _____

LEARNING ACTIVITIES

1. Pick a vitamin from the list in the book and brainstorm what kind of symptoms would result from a deficit of that vitamin. Use the Internet to check your hypothesis.

2. Polypharmacy is a real problem for many older patients. Pick a combination of drugs that might be taken together, say, cholesterol meds, calcium supplements, and blood pressure medication. Use the Internet to find out the potential interactions between the drugs. Start with two drugs, then go to three or even four. Does it get harder to predict the potential interactions as more drugs are added to the list?

3. Examine your lifestyle choices. How healthy are you? Do you engage in risky behavior? Do you get enough exercise? Eat right? How might your lifestyle choices today influence your health 40 years from now?

4. Some environmental factors are known to increase the risk of developing certain cancers. List the ones you can off the top of your head and from the textbook. Then do some research. Which cancers are influenced by environmental factors? Which factors?

5. Bacteria have increasingly become resistant to antibiotics. What factors contribute to antibiotics resistance?

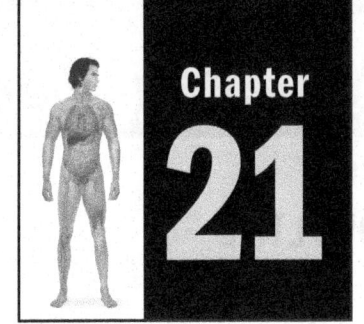

Chapter

21

HEALTH CARE: CAREERS AND CAREER PLANNING

CHAPTER SUMMARY

Most people will spend half their life or more working. So, choosing a career that will bring you satisfaction makes good sense. How will you know what will bring satisfaction? One way is to choose a career that gives you a chance to use and develop your interests, values, and abilities. Your interests are what you like to do. Your values are the elements in your life on which you place importance. Your abilities are the things you do well.

There are scores of job opportunities in the health care field, and this chapter describes many of them, along with the education, skills, and licensing required to pursue them. You can research the educational requirements, as well as other details about a selected career, by using resources in libraries and career centers, and online job and career search Web sites. Potential sources of job information also include help wanted ads, school counselors and bulletin boards, employment agencies, and your network of friends and relatives.

Developing a career plan can help you achieve your career goals. A career plan is a tool that helps you identify your skills and interests and determine the type of education and training you will need to succeed. As you progress through your career, you should take advantage of professional development opportunities, such as membership in professional associations, employer-sponsored training programs, courses at local colleges, job shadowing, and volunteering. For students, membership in a career and technical student organization can offer many benefits for developing and strengthening your career-related skills.

Once you have landed a job, it is important to uphold professional standards of appearance and behavior. Likewise, health care workers are expected to exhibit certain personal characteristics that demonstrate their commitment to the interest and welfare of their patients. They must develop effective communication skills and be aware of their ethical roles and responsibilities. These include maintaining confidentiality and being able to identify conduct and situations that may adversely affect patients, other workers, and the organization as a whole.

CHAPTER OUTLINE

I.   Identifying a suitable career
   A. Interests
   B. Values
   C. Abilities

II. Health care education and occupational resources
   A. Post-secondary options
     1. Certification
     2. Associate's degree
     3. Bachelor's degree
     4. Master's degree
     5. Doctoral degree
   B. Resources
     1. Government, employer, and job search Web sites
     2. Libraries
     3. Career centers and employment agencies

III. Professional and personal characteristics
   A. Standards of appearance
   B. Standards of behavior
   C. Personal characteristics
   D. Effective communication skills

IV. Ethical roles and responsibilities
   A. Recognizing illegal and unethical behavior
   B. Reporting illegal and unethical behavior

V.   Meetings in the Workplace
   A. Types of meetings
   B. Leading an effective meeting

VI. Safety in the workplace
   A. Understanding employee rights
   B. Understanding employee responsibilities

VII. Job-seeking skills and professional development
   A. Career portfolio
   B. Career plan
   C. Professional development
   D. Career and technical student organizations
   E. Volunteering

MULTIPLE CHOICE

Circle the letter of the correct answer.

1. Which type of doctor diagnoses, treats, and tries to prevent problems of the heart and blood?
   a. urologist
   b. chiropractor
   c. cardiologist
   d. none of the above

2. When a patient's kidneys are not functioning properly, he must undergo a process to have waste products filtered from the blood. The primary caregiver for the patient is a(n):
   a. dialysis technician.
   b. dietician.
   c. respiratory therapist.
   d. ECG/EKG technician.

3. Which type of physician treats conditions and diseases of the skin, hair, and nails?
   a. endocrinologist
   b. epidemiologist
   c. hematologist
   d. dermatologist

4. Which type of physician specializes in treatment of the liver?
   a. immunologist
   b. epidemiologist
   c. nephrologist
   d. hepatologist

5. Which type of doctor would treat someone with allergies or autoimmune disease?
   a. immunologist
   b. internist
   c. gene therapist
   d. oncologist

6. Which type of doctor specializes in the diagnosis and treatment of mental illness?
   a. pulmonologist
   b. radiologist
   c. psychiatrist
   d. endocrinologist

7. In which health care career would you work with disabled people to help them maximize their abilities to perform the tasks of daily life?
   a. internist
   b. occupational therapist
   c. orthopedist
   d. podiatrist

8. Which type of doctor uses medical imaging techniques to diagnose and treat disease?
   a. pulmonologist
   b. radiologist
   c. urologist
   d. proctologist

9. Which career would you choose if you wanted to study human behavior and the human mind?
   a. physical therapist
   b. respiratory therapist
   c. pathologist
   d. psychologist

10. Which career would you choose if you wanted to specialize in health care for infants, children, and teenagers?
    a. otologist
    b. ophthalmologist
    c. pediatrician
    d. pathologist

11. Which type of physician specializes in treating diseases and conditions of the chest, including the lungs and bronchial tubes?
    a. pulmonologist
    b. radiologist
    c. nephrologist
    d. hematologist

12. Which career would you choose if you wanted to collect and analyze physical evidence that could be used in solving a crime?
    a. forensic scientist
    b. epidemiologist
    c. pathologist
    d. hematologist

13. Which career would you choose if you wanted to study the occurrence and frequency of disease in large populations?
    a. endocrinologist
    b. epidemiologist
    c. cardiologist
    d. urologist

14. Which career would you choose if you wanted to design equipment and supplies that help solve medical and health-related problems?
    a. physical therapist
    b. occupational therapist
    c. forensic scientist
    d. biomedical engineer

15. Which type of physician specializes in nervous system disorders?
    a. obstetrician
    b. ophthalmologist
    c. neurologist
    d. nephrologist

MATCHING EXERCISES

Set 1

Please match the description with the appropriate job title.

_____ 1. Health care worker who is involved in every aspect of patient care, including diagnostic testing and analysis, giving treatment and medication, observing patients and recording progress, and follow-up and rehabilitation.

_____ 2. Health care worker who provides pre-hospital care in an emergency.

_____ 3. Health care worker who provides preventive care services, diagnostic services, and therapeutic services under the supervision of a physician or surgeon.

_____ 4. Health care professional who provides care and support to women throughout normal pregnancies.

_____ 5. Health care worker who provides basic bedside care for patients.

_____ 6. Health care worker who helps patients perform day-to-day tasks such as bathing, eating, and dressing; take vital signs; transports patients; and sets up equipment.

_____ 7. A registered nurse who has received additional education and clinical training in the diagnosis and treatment of illness.

a. Emergency medical technician

b. Licensed practical nurse

c. Midwife

d. Nurse practitioner

e. Nursing assistant

f. Physician's assistant

g. Registered nurse

Set 2

Please match the description with the correct personal characteristic.

_____ 1. Having the ability to do something well, measured against a standard, especially ability acquired through experience or training.

_____ 2. Being fair, truthful, and morally upright.

_____ 3. Being able to deal with disapproval or a suggestion that something can be improved.

_____ 4. Being sensitive to and experiencing the feelings, thoughts, and experiences of another.

_____ 5. Showing good judgment and sensitivity needed to avoid embarrassing or upsetting others.

_____ 6. Being able to admit that you don't know the answer or that you can accept help to understand a situation more fully.

_____ 7. Being accountable to somebody or for something.

_____ 8. Working cooperatively with others and subordinating personal interests in order to achieve a common goal.

_____ 9. Having the ability to put up with waiting, delay, or provocation without becoming annoyed or upset, or to act calmly when faced with difficulties.

_____ 10. Being energetic, ambitious, and able to get things done without being directed by others.

_____ 11. Showing excited interest in or eagerness to do something.

_____ 12. Being reliable and trustworthy.

a. Acceptance of criticism

b. Competence

c. Dependability

d. Discretion

e. Empathy

f. Enthusiasm

g. Honesty

h. Patience

i. Responsibility

j. Self-motivation

k. Teamwork

l. Willingness to learn

FILL IN THE BLANK

Fill in the blanks to complete the following statements.

1. A(n) _____ is something you like to do.

2. A(n) _____ is the importance that you place on various elements in your life.

3. A(n) _____ is something you do well.

4. A(n) _____ degree is usually awarded after completion of a two-year college degree.

5. A(n) _____ degree is usually awarded after completion of a four-year college degree.

6. A code of conduct representing ideal behavior for a group of people is referred to as _____.

7. The _____ requires all employers to provide a safe and healthful workplace.

8. A(n) _____ is a legally binding document that ensures that the medical facility will protect patient information.

9. The _____ was formed by the federal government to inspect companies and enforce safety laws.

10. A(n) _____ is a group that promotes career education and training and provides opportunities to develop leadership, teamwork, and citizenship skills.

11. A(n) _____ is a collection of materials that show the knowledge, abilities, skills, and insights you gain in your search for a career.

12. A(n) _____ is a strategy for a person's growth as a professional.

13. Doing work without pay is referred to as _____.

14. A(n) _____ is a process by which you examine your strengths and weaknesses to learn more about your career-related skills, interests, values, and abilities.

15. Spending a period of time with an experienced worker and observe his job-related duties and responsibilities is referred to as _____.

SHORT ANSWER

1. List five standards of appearance for all health care workers.

    _____

    _____

    _____

2. List at least five standards of behavior for all health care workers.

    _____

    _____

    _____

3. List the steps you should follow for effective communication.

    _____

    _____

    _____

4. List and explain the rights and responsibilities set forth for every employee by the Occupational Safety and Health Act.

    _____

    _____

    _____

5. List and explain at least three benefits of being a member in a CTSO.

    _____

    _____

    _____

    _____

LEARNING ACTIVITIES

1. Sometimes it is difficult to identify your own interests, values, and abilities. Working in pairs, spend 15 minutes talking to your partner about the things you like to do and the things that are important to you. Discuss the things you think you are good at, your favorite subjects in school, and the clubs or organizations you have joined. Take notes so you will remember what your partner tells you.

    When you have finished talking, look over your notes and select a career from the jobs and professions described in this text or from other resources and select a career that you think suits your partner's interests, values, and abilities. Write a letter of recommendation to a potential employer on behalf of your partner, explaining why you think this would be the right career match.

2. From the list of jobs and professions discussed in this chapter, select one that interests you. Research the job using the *Occupational Outlook Handbook*, the Dictionary of Occupational Titles, and other resources. Prepare a job description that outlines the duties and responsibilities of the job, the education and training requirements, salary range, working environment, and employment outlook.

3. Working in teams of three to five, research and gather information on workplace code of ethics. You should find information on the purpose of a code of ethics and the issues and topics that a code addresses. Using the information you find, create a poster that outlines a general workplace code of ethics.

 ## CASE STUDY

Samantha is an emergency room health care employee who has witnessed her co-worker, Jason, take a personal call on his cell phone while questioning a patient who has just come in complaining of stomach pain.

1. Do you think Jason taking the personal call is a reportable behavior? Why or why not?

_____

_____

_____

_____

2. Can you identify the problem or challenge facing Samantha?

_____

_____

_____

_____

3. What do you think Samantha should do next? Explain your answer.

_____

_____

_____

_____